CALCULATIONS IN FOOD AND CHEMICAL ENGINEERING

Theory, worked examples and problems

A.T. Jackson
Department of Chemical Engineering,
University of Leeds

J. Lamb
Procter Department of Food Science,
University of Leeds

First published 1981 by
THE MACMILLAN PRESS LTD
London and Basingstoke
Companies and representatives
throughout the world

Printed in Hong Kong

ISBN 0 333 29423 8

CONTENTS

PREFACE

This book is intended primarily for students taking chemical engineering and food technology and engineering courses and aims to show procedures which can be used in process design. Each section deals with a particular unit operation, giving a brief description of the relevant theory underlying the operation and a number of worked and unworked examples to show how the theory is applied. The processes have been selected with particular reference to the food industry, but the treatment is also relevant to other process industries.

Space does not allow a comprehensive treatment of the theory but references are given to guide the reader in further reading. It is recognised by the authors that engineers working day-by-day on process analysis of particular operations have developed their own procedures, many of which are not published. These procedures may be different from those given in this text, but it is hoped that the treatment given here will act as a stimulus for further consideration of the most suitable methods of process analysis.

The nomenclature used is defined within the text in each chapter. The S.I. system of units is used throughout. One particular unit which in the authors' experience causes difficulties for students is that for viscosity: three commonly used units for dynamic viscosity are all used in the text - pascal second (Pa s), poise (P) and kg/m s, where 1P = 0.1 Pa s = 0.1 kg/m s. Similarly pressure has been defined by both N/m^2 and Pa, where 1 Pa = $1N/m^2$.

In the worked examples we have retained a large number of significant figures within the calculation, but the final figure in the example answer has been rounded off to a reasonable degree of precision.

Design equations to quantify the effect of process variables (for example in calculation of process time) are frequently based on empirical correlations. This may lead to errors when they are applied to different conditions from those from which the correlations were established. However the equations may well give a good estimate of the effect of a change in a process variable, even though the absolute values may be inaccurate. For example, although heat transfer correlations are unlikely to allow prediction of heat transfer coefficients to within better than 15% accuracy, a knowledge that heat transfer coefficients vary with $(\text{flow velocity})^{0.8}$ in turbulent flow can be very useful in design. Examples of this are given in the text.

Almost inevitably there will be some errors in the worked calculations and we would be pleased to hear of these from readers, together with any suggestions to improve the usefulness of the book.

We acknowledge with gratitude the permission to publish figures 2.6, 2.7, 2.8, 6.1 (John Wiley and Sons), 2.12 (Elsevier Publishing Co Co. Ltd.,) 4.12 (Pergamon Press). Our thanks also go to Noreen Green for her skilled assistance in preparing the typescript.

J. Lamb

A. T. Jackson

1 FLUID TRANSPORT

1.1 BASIC FLUID FLOW

NEWTONIAN FLOW

A Newtonian fluid is defined as a fluid flowing over a surface exhibiting a constant value for viscosity (μ) for all conditions of shear stress (R) and rate of shear (du/dy) in the expression

$$R = -\mu(du/dy)$$

u = longitudinal velocity of fluid; (du/dy) = velocity gradient measured perpendicular to surface. Thus, a plot of R against (du/dy) will be a straight line of slope $-\mu$ through the point (0,0).

For the flow of a Newtonian fluid through a pipe or channel, the head loss due to friction can be shown to be

$$\Delta H_f = 2f(L/D_e)(u^2/g) \quad \text{or} \quad 4\Phi(L/D_e)(u^2/g)$$

L= length of pipe or channel; u = average velocity; f = Fanning friction factor; Φ = Stanton and Pannel [1] friction factor; D_e = 'equivalent diameter' of the channel = 4(cross-sectional area/wetted perimeter). Both friction factors are a function of the Reynolds number ($D_e u\rho/\mu$) [2] and the surface relative roughness (ξ/D_e), where ρ = fluid density.

Laminar (Streamline) Flow [Re<2100]

The friction factor in this flow region is independent of surface roughness and for circular pipes only

$$f = 16/Re \quad \text{or} \quad \Phi = 8/Re \quad \text{and Re = Reynolds number.}$$

For non-circular sections precise solutions are available [1,4].

Turbulent Flow [Re>2100]

For smooth pipes in the range $2500<Re<10^5$, the Blasius equation [3] can be used

$$\Phi = \tfrac{1}{2}f = 0.0396(Re)^{-0.25}$$

For rough pipes, the expression due to Moody [4] may be used for the range $3000<Re<10^7$

$$\Phi^{-\frac{1}{2}} = -2.5\ \ln[0.27(\xi/D_e) + 0.885\Phi^{-\frac{1}{2}}/Re]$$

Alternatively, graphs of friction factor versus Reynolds number can be found in most textbooks, similar to figure 1.1.

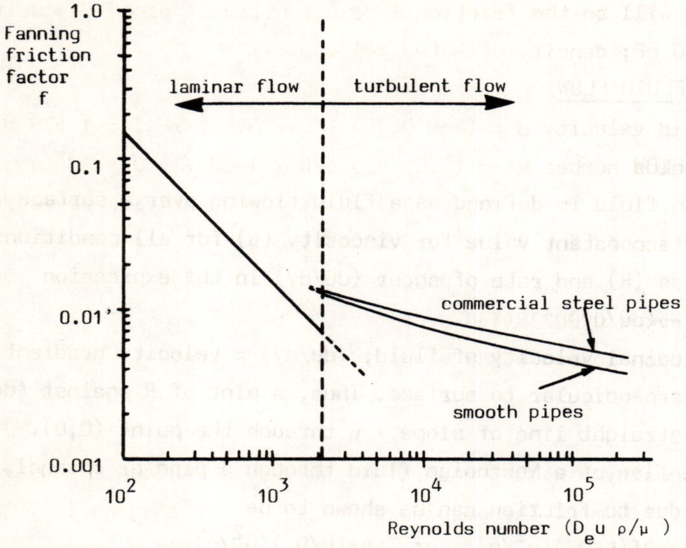

Fig 1.1 - Fanning friction factor against Reynolds number

For pipe fittings, the friction loss is expressed in a number of ways

(a) as a number of 'velocity heads' lost; velocity head = $u^2/2g$

(b) as an 'equivalent length' of pipe or channel the same size as the fitting

(c) as an equivalent length of pipe diameters; similar in effect to (b) above, but requiring less tabulated data.

Fitting	Equivalent Length	Equivalent Diameters
1 in. Globe valve - open	28 ft.	336
1½ in. Globe valve - open	42 ft.	336
1½ in. bend	4½ ft.	36
2 in. bend	6 ft.	36

The equivalent length (L_e) of the fitting is added to the overall length of straight pipe in the expression for ΔH_f. The velocity heads lost are added to the head loss for the straight pipe calculated separately.

Example 1.1

0.18 m^3/minute of water is flowing in a smooth pipe of 50 mm nominal bore. What will be the friction loss per metre of pipe? Viscosity of water = 1.0 cP; density of water = 1000 kg/m^3.

The fluid velocity $u = (4 \times 0.18)/(60 \times 0.05^2 \times \pi) = 1.528$ m/s

Reynolds number $Re = (0.05 \times 1.528 \times 1000)/(1.0 \times 10^{-3})$

$= 76\ 349$ i.e. turbulent flow region.

Friction factor $\Phi = \frac{1}{2}f = 0.0396(76\ 349)^{-0.25}$

$= 0.0396 \times 0.0602 = 0.00238$

$\Delta H_f = 4 \times 0.00238(1.0/0.05)(1.528^2/9.81)$

Friction loss = 0.045 m water/meter of pipe.

Example 1.2

Oil of viscosity 5×10^{-3} kg/m s, is flowing in an annulus with a velocity of 1.0 m/s. The outside diameter of the annulus is 75 mm, the inside diameter is 50 mm. If the annulus is 100 m long, what will be the friction loss over the annular section? The density of the oil = 850 kg/m^3.

The equivalent diameter of the annulus is given by

D_e = (4 x cross-sectional area)/(wetted perimeter)

$= 4\pi(D_o^2 - D_i^2)/[\pi(D_o + D_i)] = (D_o - D_i)$

D_o = outside diameter; D_i = inside diameter

Reynolds number $Re = (0.075 - 0.05) \times 1.0 \times 850/(5 \times 10^{-3})$

$= 4250$ i.e. turbulent flow region.

Friction factor $\Phi = \frac{1}{2}f = 0.0396(4250)^{-0.25}$

$= 0.0396 \times 0.1239 = 0.0049$

$\Delta H_f = (4 \times 0.0049 \times 100 \times 1.0^2)/[(0.075 - 0.05) \times 9.81]$

$= 8.0$ m head of oil.

Example 1.3

An open-top channel is used to convey treated effluent. The length of the channel is 1500m, and the channel is 200 mm wide x 150 mm deep. The effluent is found to run 100 mm deep at a flowrate of 3.0 m^3/h. What will be the friction loss in the channel ? Effluent viscosity = 2×10^{-3} kg/m s; effluent density = 1080 kg/m^3.

Cross sectional flow area = $0.2 \times 0.1 = 0.02\ m^2$

Wetted perimeter = $(0.2 + 0.1 + 0.1) = 0.4$ m

Equivalent diameter $D_e = (4 \times 0.02)/0.4 = 0.2$ m

Velocity of flow = $3.0/(3600 \times 0.02) = 0.042$ m/s

Reynolds number $Re = (0.2 \times 0.042 \times 1080)/(2 \times 10^{-3})$

$= 4536$ i.e. turbulent flow region.

$\Phi = \frac{1}{2}f = 0.0396(4536)^{-0.25} = 0.0048$

$\Delta H_f = 4 \times 0.0048(1500/0.2)(0.042^2/9.81) = 0.026$ m head of effluent.

Example 1.4

A piping system consists of 200 m of 50 mm bore pipe, and contains 5 bends. The end of the pipe system is connected to a globe valve before entering a storage tank. What will be the total friction loss in the system for a flowrate of $0.58\ m^3/h$? The fluid viscosity $= 1.5 \times 10^{-3}$ kg/m s; fluid density = $1100\ kg/m^3$. Each bend is equivalent to 0.8 velocity head, and the globe valve is equivalent to 336 diameters.

Fluid velocity = $(4 \times 0.58)/(3600 \times \pi \times 0.05^2) = 0.082$ m/s

Reynolds number $Re = (0.05 \times 0.082 \times 1100)/1.5 \times 10^{-3})$

$= 3008$ i.e. turbulent flow region.

$\Phi = \frac{1}{2}f = 0.0396(3008)^{-0.25} = 0.00535$ (assuming a smooth pipe)

The length of straight pipe is 200 m, but this must be increased by the equivalent length of the fittings.

Globe valve equivalent diameters = 336

Equivalent length of pipe L_e, 50 mm bore = (336×0.05) = 16.8 m

The bends have a resistance equivalent to 0.8 velocity heads each, and entry losses into the tank 1 velocity head.

Total velocity heads lost = $1 + (5 \times 0.8) = 5.0$

Using this value multiplied by the velocity head $(u^2/2g)$ will give the friction loss for these items directly in head of fluid.

$H_1 = (0.082^2 \times 5.0)/(2 \times 9.81) = 0.00171$ m

H_1 can be added to the friction losses due to the straight pipe and other fittings, H_2, where

$H_2 = 4 \times 0.00535 \times (200 + 16.8) \times 0.082^2 /(0.05 \times 9.81) = 0.0636$ m

Therefore, the total friction losses, $\Delta H_f = H_1 + H_2$

$\Delta H_f = 0.00171 + 0.0636 = 0.065$ m of fluid flowing.

TIME-INDEPENDENT NON-NEWTONIAN FLUIDS

With a non-Newtonian fluid, a plot of shear stress (R) against shear rate (du/dy) is not linear. Such fluids are commonly classified as follows.

Bingham plastics. A certain 'yield stress' (R_y) must be applied before flow starts, giving a relationship

$$(R - R_y) = - \mu_p (du/dy)$$

μ_p = 'plastic viscosity' or 'rigidity', and $R > R_y$.

Pseudoplastics. The ratio of R to (du/dy) - the apparent viscosity (μ_a)-falls with increasing shear rate. The flow behaviour of many pseudoplastic fluids may be described by a simple power law relationship

$$R = K(du/dy)^n$$

K = 'fluid consistency'; n = 'flow behaviour index' or 'power law index' = <1.0.

Dilatant fluids. The apparent viscosity increases with increasing shear rate. The power law often applies to this type of fluid, but $n > 1.0$.

Where the fluid does not follow the power law, the following relationship is used to describe the fluid characteristics, but applies for point values of the shear stress and shear rate

$$R = K'(du/dy)^{n'}$$

K' = fluid consistency; n' = flow behaviour index or 'generalised rheological parameter'.

Fluids in Laminar Pipe Flow

Since most non-Newtonian fluids of interest in the food, chemical and processing industries exhibit high values for the apparent viscosity, it is unusual to find fully developed, turbulent flow. Hence, only the laminar flow region will be treated in any detail.

Based on the work of Rabinowitsch [5] and Mooney [6], it was shown by Metzner and Reed [7] that for <u>laminar flow</u>, the shear stress at the wall of a pipe (R_w) is given by

$$R_w = (D\,\Delta P/4L) = K'(8u_m/D)^{n'}$$

D = pipe diameter; L = pipe length; ΔP = pressure drop over length L; u_m = mean fluid velocity = (flowrate/cross-sectional area); K' = fluid consistency; n' = flow behaviour index.

This expression can be used with any fluid, (Bingham plastic, pseudoplastic, dilatant or Newtonian), and for Newtonian fluids, n' = 1.0, K' = μ and the expression reduces to the familiar Poiseuille equation.

Thus, a log/log plot of the values of (D ΔP/4L) versus ($8u_m$/D) gained from an experimental rig using at least two pipe sizes, will allow determination of the fluid characteristics.

(a) If the plot is a straight line, the power law applies and K and n can be evaluated for use in the expression for pressure drop over the range of experimental shear rate values.

(b) If the plot is not a straight line, provided that the range of values of ($8u_m$/D) obtained experimentally correspond to the values expected on a full scale plant, the data can be used satisfactorily to design the piping system.

<u>Example 1.5</u>

A thick, particulate suspension is investigated in the laboratory by pumping it at different rates through a series of pipes and measuring the pressure drop. The results obtained are tabulated below.

D ΔP/4L (N/m^2)	13.36	14.40	15.23	16.44	19.17	21.17	22.53
$8u_m$/D (s^{-1})	5.5	8.0	0.6	15.5	33.5	55.0	75.0

What information regarding the characteristics of this suspension can be deduced from these results ?

A log/log plot of (D ΔP/4L) versus ($8u_m$/D) gives a straight line relationship having a slope of 0.2.

It can be deduced from this that

(a) the fluid is pseudoplastic in behaviour and follows the power

(b) over the range of shear stresses and shear rates investigated, the fluid follows the relationship

$$(D\,\Delta P/4L) = K(du/dy)^n$$

The flow behaviour index n = 0.2, and by taking values of $(D\,\Delta P/4L)$ and $(8u_m/D)$, the mean value for the fluid consistency K= 9.50. Thus, the generalised fluid flow relationship for this suspension is

$$(D\,\Delta P/4L) = 9.50(du/dy)^{0.2}$$

Example 1.6

The thick, particulate suspension in example 1.5 is to be pumped to a process vessel at a rate of 0.2 m^3/min using an existing pump capable of developing a pressure of 2.0 bar. The equivalent length of pipework and fittings is 200 m. What will be the smallest possible pipe size which can be used ?

Mean fluid velocity $u_m = (0.2 \times 4)/(60 \times \pi \times D^2) = 0.00424/D^2$

Hence $(8u_m/D) = (8 \times 0.00424)/D^3 = 0.03392/D^3$.

Substituting in the generalised flow expression

$$(D \times 2 \times 10^5)/(4 \times 200) = 9.50(0.03392/D^3)^{0.2}$$

$$D^{1.6} = (9.50 \times 4 \times 200 \times 0.03392^{0.2})/(2 \times 10^5)$$

$$= 0.038 \times 0.5083 = 0.01932$$

Therefore D = 0.0849 m i.e. 85 mm bore minimum diameter

It is interesting to note that for highly non-Newtonian pseudoplastic fluids such as the one treated in examples 1.5 and 1.6, the pressure drop (ΔP) shows the following relationship [8]:

$$\Delta P \propto Q^{n'}/D^{(3n'+1)}$$

Thus, when the fluid is highly non-Newtonian, i.e. when n' is close to zero, the effect of throughput (Q) on ΔP is small compared to the effect of pipe size (D). A substantial increase in flowrate for such fluids can often be achieved in the same piping system without too great an increase in pressure drop. This is not the case with Newtonian fluids where n' = 1.

An alternative method suggested by Metzner and Reed is in the use of standard Newtonian friction factor charts or correlations using a modified Reynolds number (Re')

$$Re' = (D^{n'} u_m^{2 - n'} \rho)/(K' 8^{n' - 1})$$

and, for laminar flow

$$\Phi = 8/Re' \quad \text{or} \quad f = 16/Re'$$

can be used in the expression

$$\Delta H_f = 4\Phi(L/D)(u_m^2/g) = 2f(L/D)(u_m^2/g)$$

Example 1.7

For the thick suspension in example 1.6 what will be the pressure drop in the system using 85 mm bore piping? The density of the suspension is 1240 kg/m^3.

The modified Reynolds number $Re' = (D^{n'} u_m^{2 - n'} \rho)/(K' 8^{n' -1})$. From example 1.6, $u_m = 0.00424/D^2 = 0.587$ m/s.

$$Re' = (0.085^{0.2} \times 0.587^{1.8} \times 1240)/(9.50 \times 8^{-0.8})$$
$$= (0.6108 \times 0.3833 \times 1240)/(9.50 \times 0.1895) = 161.3$$

The Fanning friction factor f = 16/161.3 = 0.0992

$$\Delta H_f = 2 \times 0.0992(200/0.085)(0.587^2/9.81) = 16.4 \text{ m of suspension.}$$

But the pressure drop

$$\Delta P = \rho g \Delta H_f = 1240 \times 9.81 \times 16.4 = 199\,496 \text{ N/m}^2 = 1.99 \text{ bar,}$$

which compares favourably with the specified pressure drop in example 1.6.

Fluids in Turbulent Pipe Flow

Dodge and Metzner [9] showed that for pseudoplastic fluids obeying the power law, the onset of turbulent flow occurs at modified Reynolds numbers which increase as the power law index (n) decreases, and a log/log plot of ($D\Delta P/4L$) against ($8u_m/D$) gives a series of curves in the turbulent region for each pipe diameter.

Dodge and Metzner suggested the use of a modified Blasius equation to obtain friction factor values

$f = a(Re')^b$ where, a and b are functions of the flow behaviour index n'.

The results of Dodge and Metzner and Bowen [10] can be correlated for the range $0.2 < n' < 2.0$ as

$$a = 0.078(n')^{0.11} \quad \text{and} \quad b = 0.252(n')^{-0.211}$$

and may be used in the modified Blasius equation to give values of friction factor (f) for use in the conventional head loss expressions.

1.2 PUMP SIZING

CENTRIFUGAL PUMPS

In the selection of a centrifugal pump for any specific duty, the main characteristics required to specify the pump are

(a) the rate of discharge

(b) the head to be developed by the pump.

For the specification of a new pump to perform a certain duty, the rate of discharge is dictated by process conditions, and the head to be developed is a function of the rate of flow and the piping system.

An alternative industrial problem is the case where a pump already exists (perhaps from a discontinued process), and it is required to determine if the pump can be used for another duty.

The head to be developed by any pump can be expressed as follows

$$\Delta H = H_s + H_f$$

ΔH = total head required; H_s = static head; H_f = head losses due to friction in the piping system.

The frictional losses will be due to a combination of fluid flow in the straight sections of pipe, and also the losses due to fluid flow through valves, bends and other pipe fittings.

For the specification of a new pump, it is sufficient to calculate the total head to be developed (ΔH), and then compare the characteristic curves of several pumps to choose one giving the specified flowrate at the calculated head.

In the case where it is desired to asses the suitability of an existing pump, since the flowrate which will be achieved is unknown, it becomes necessary to develop a 'characteristic duty curve' for the <u>system</u> for comparison with the characteristic curve of the <u>pump</u>.

Example 1.8

It is required to pump milk from a storage tank located at ground level to a process vessel situated 10.0 m above ground level. The pipeline from the storage tank to the process vessel is 200 m long x

75 mm bore. The entry losses into the process vessel and through associated valves and fittings are equivalent to 10 velocity heads (based on the flow in the 75 mm pipe).

A centrifugal pump is available with the following characteristics

Flowrate (m^3/h)	10.0	14.26	18.34	20.38	21.4
Head (m)	23.0	21.43	18.9	15.24	11.0

Density of milk = 1150 kg/m^3; viscosity = 2×10^{-3} kg/m s. Calculate the expected rate of flow and the power required if the pump system is 55% efficient.

The head which must be developed by the pump is

$$\Delta H = H_s + 2f(L/D)(u^2/g) + 10.0u^2/2g$$

The minimum value of the Reynolds number will give the maximum value for the Fanning friction factor (f), and will represent the most pessimistic case. The minimum flowrate given in the pump specification is 10.0 m^3/h.

$$\text{Minimum Reynolds number } Re_m = (4Q\rho)/(\pi D\mu)$$

$$= (4 \times 10 \times 1150)/(\pi \times 0.075 \times 2 \times 10^{-3} \times 3600)$$

$$= 27\ 100$$

Based on this value

$$f = 0.0792(Re_m)^{-0.25} = 0.0062$$

therefore

$$\Delta H = 10.0 + u^2(10/2 \times 9.81) + (2 \times 0.0062)(200/0.075)(u^2/9.81)$$

$$= 10.0 + 3.371u^2$$

$$\text{Discharge rate } Q = (\pi D^2 u)/4$$

and $u = 4Q/(\pi D^2) = 0.063Q'$ where Q' is in m^3/h.

therefore

$$\Delta H = 10.0 + 0.0133Q'^2$$

Using this expression, values of the head to be developed by the pump for this particular system can be calculated for various discharge rates.

By plotting these results on the same axes as the pump characteristic curve, the point of intersection of the two curves gives the expected rate of flow. From figure 1.2, the expected rate of flow is 20.2 m^3/h, at a head of 15.5 m.

Theoretical power required = $mg\Delta H$

where m = mass flowrate, but since the pump system is only 55%

Q' (m^3/h)	ΔH (m of milk)
10	11.33
12	11.92
14	12.61
16	13.40
18	14.31
20	15.32
22	16.44

efficient, the actual power required will be

$$(20.2 \times 1150 \times 9.81 \times 15.5)/(3600 \times 0.55) = 1784 \text{ W} = 1.78 \text{ kW}.$$

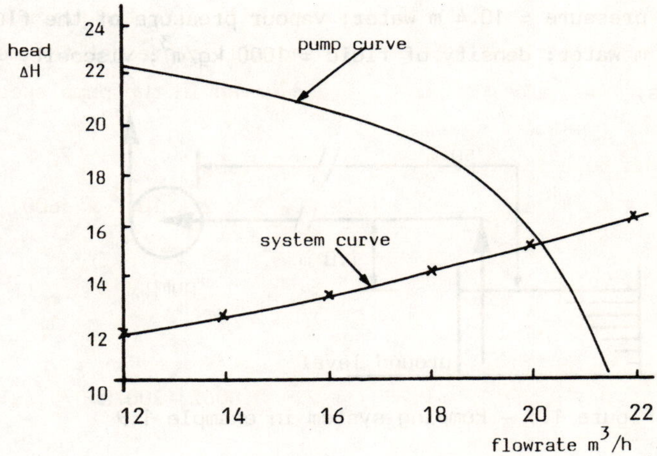

Figure 1.2 - Plot of head (ΔH) against flowrate (Q), example 1.8

Another important factor in the use of centrifugal pumps is the consideration of conditions at the pump suction. If the suction head is high, the fluid in the suction port can vaporise, followed by condensation in either the pump impeller or casing (cavitation). Since the mechanical shocks due to cavitation are detrimental to the pump, this situation should be avoided. All pump manufacturers quote a 'Net Positive Suction Head' (NPSH) which they recommend as the minimum head to be maintained at the pump suction to avoid the risk of cavitation

$$NPSH = (P_o/\rho g) - (P_v/\rho g) - H_f - (u^2/2g) + H_s$$

P_o = atmospheric pressure; P_v = vapour pressure of the fluid at the temperature of operation; H_s = static head at pump suction; H_f = friction losses in the pump suction piping; $(u^2/2g)$ = velocity head loss at the immediate entry to the impeller.

Example 1.9

Centrifugal pumps are available from a number of manufacturers, each capable of delivering 12.0 m^3/h at 40 m head.

Pump 1	has a 32 mm	bore suction branch,	40 mm bore delivery branch
2	40 mm		40 mm
3	50 mm		40 mm

Which would be the most suitable pump for use in the system shown in figure 1.3, if the NPSH is quoted at 1.5 m of water in all three cases? Atmospheric pressure = 10.4 m water; vapour pressure of the fluid at 20'C = 1.23 m water; density of fluid = 1000 kg/m^3; viscosity = 0.001 kg/m s.

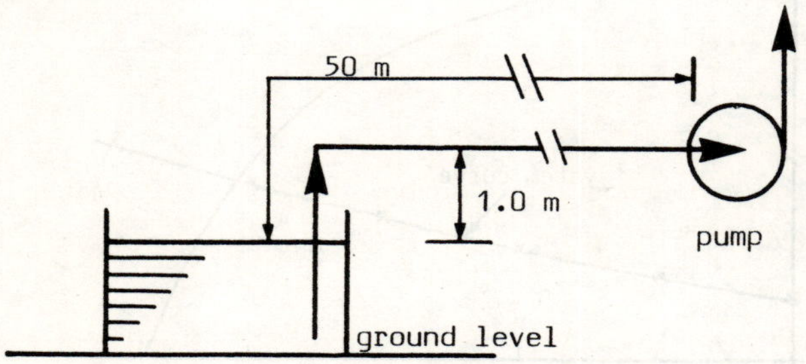

Figure 1.3 - Pumping system in example 1.9

$P_o/\rho g$ = 10.4; $P_v/\rho g$ = 1.23; H_s = -1.0 (feed below suction);

$H_f = 2f(L/D)(u^2/g)$.

$$NPSH = 10.4 - 1.23 - 1.0 - 2f(50/D)(u^2/g) - (u^2/2g)$$
$$= 8.17 - (u^2/g)(2f \times 50/D + 0.5)$$

The Fanning friction factor (f) is a function of the Reynolds number in the suction piping. Since the flowrate in each case is 12.0 m^3/h

Fluid velocity $u = (4 \times 12)/(\pi \times D^2 \times 3600) = (4.25 \times 10^{-3})/D^2$ m/s

Assuming that the suction piping will be the same diameter as each pump suction branch, the following table can be drawn up

D (m)	u (m/s)	Re (Duρ/μ)	f	NPSH (m)
0.032	4.15	1.33×10^5	0.004	-14.65
0.040	2.66	1.06×10^5	0.0044	-0.125
0.050	1.70	8.50×10^5	0.0045	+5.37

Since the minimum NPSH is quoted at 1.5 m water, pump number 3 is the only one suitable for this duty. (It would be possible to assess pump number 2 using 50 mm bore suction piping to reduce frictional losses - this gives a value for the NPSH of + 5.16 m, and would also be suitable).

POSITIVE DISPLACEMENT PUMPS

In the case of positive displacement pumps operating with incompressible fluids, the average discharge rate remains constant, irrespective of the head developed (and hence pressure), and discharge conditions are not of over-riding importance, except that the discharge piping must be protected against excessive pressures.

However, conditions at the pump inlet are important if vaporisation of the liquid is to be avoided, and the speed of the pump may have to be closely regulated to avoid 'separation' (vaporisation) in the suction piping or body of the pump.

Example 1.10

A single-acting reciprocating pump has a cylinder diameter 115 mm and a stroke of 250 mm. The piston can be considered to move with simple harmonic motion.

The suction line is 10.0 m long x 50 mm bore, and the feed tank is located 2.0 m below the pump cylinder. If separation occurs at an absolute head of water of 1.20 m, what will be the maximum speed of the pump and maximum output? Atmospheric pressure = 10.4 m water; density of fluid = 1100 kg/m^3; viscosity = 1.4 x 10^{-3} kg/m s.

The greatest tendency for separation to occur will be at the start of the suction stroke and at the inlet to the cylinder.

If the piston is moving with simple harmonic motion, the acceleration (a) of the piston is given by

$$a = r\omega^2\cos(\omega\theta)$$

r = half the stroke; ω = angular velocity, rad/s; θ = time.

The maximum acceleration will occur when θ = 0, and

$$a_{max} = r\omega^2$$

Transferring this cylinder acceleration to the suction line maximum acceleration (a'_{max})

$$a'_{max} = (115/50)^2 r\omega^2$$

if N rev/s = speed of rotation of the drive, $\omega = 2\pi N$ rad/s

$$a'_{max} = (115/50)^2 (0.250/2)(2\pi N)^2 = 26.11N^2$$

Accelerating force on the liquid (F) is given by

$$F = (\text{mass of liquid} \times \text{acceleration}) = AL\rho a'_{max}$$

A = cross-sectional area of piping; L = length of piping.

$$\text{Acceleration head} = (F/A\rho g) = La'_{max}/g = 10.0 \times 26.11N^2/9.81$$
$$= 26.6N^2 \text{ m of liquid}$$

Separation occurs at 1.20 m water = 1.20(1000/1100) = 1.091 m of liquid

Atmospheric pressure = 10.4 m water = 10.4(1000/1100) = 9.455 m of liquid

A head balance over the system gives

$$1.091 = 9.455 - 2.0 - 26.6N^2$$

and $N = 0.489$ rev/s

Therefore, the maximum speed of the pump = 29.3 rev/min.

Maximum output = (stroke)(rate of stroke)(cylinder cross-section)

$$= 0.250 \times 0.489 \times \pi \times (0.115^2/4)$$
$$= 1.27 \times 10^{-3} \text{ m}^3\text{/s} = 4.57 \text{ m}^3\text{/h.}$$

REFERENCES

1. T. Stanton and J. Pannel, 'Similarity of motion in relation to the surface friction of fluids', Phil.Trans.R.Soc., 214 (1914) 199

2. O. Reynolds, 'On the dynamical theory of incompressible viscous fluids and the determination of the criteria', Phil.Trans.R.Soc., 177 (1886) 157

3. H. Blasius, 'Das Ähnlichkeitsgesetz bei Reibungsvorgangen in Flüssigkeiten', Forschungsarbeiten Ver.dt.Ing., No 131 (1913)

4. L.F. Moody, 'Friction factors for pipe flow', Trans.Am.Soc.Mech. Engrs, 66 (1944) 671

5. B. Rabinowitsch, 'Über die Viscosität und Elastizität von Solen', Z.Phys.Chem., A145 (1929) 1

6. M. Mooney, 'Explicit formulas for slip and fluidity', J.Rheol., 2 (1931) 210
7. A.B. Metzner and J.C. Reed, 'Flow of non-Newtonian fluids - correlations of the laminar, transition and turbulent-flow regions', A.I.Ch.E.Jl, 1 (1955) 434
8. W. Wilkinson, Non-Newtonian Fluids, (Pergamon, London, 1960)
9. D.W. Dodge and A.B. Metzner, 'Turbulent flow of non-Newtonian systems', A.I.Ch.E.Jl, 5 (1959) 189
10. R. Le B. Bowen, 'Designing turbulent flow systems', Chem. Engng, 68(15) (1961) 143

2 HEAT TRANSFER

2.1 STEADY-STATE CONDUCTION AND CONVECTION

CONDUCTION

Conduction is the transfer of heat from a region of higher temperature to a region of lower temperature within a medium (solid, liquid or gas), or between different materials in physical contact, without appreciable movement of the molecules of the material.

The basic relationship for conduction heat transfer was proposed by Fourier in 1882

$$dQ/d\theta = -kA(dt/dx)$$

$(dQ/d\theta)$ = rate of heat flow; (dt/dx) = temperature gradient across a thickness dx; A = area through which heat is flowing, measured normal to the direction of flow; k = thermal conductivity of the material (a basic property of the material); θ = time.

Once a steady-state has been reached (defined by the temperature at any point in the system being independent of time), the relationship for the flow of heat can be written

$$q = kA(\Delta t/x)$$

q = rate of heat transfer; Δt = temperature difference between boundaries; x = thickness of material.

Because the thermal gradient is negative, in order to be mathematically correct, the Fourier equation should include the negative sign, but in practice this is frequently omitted.

Most practical problems involve the transfer of heat through composite materials in series, and it is convenient to re-write the Fourier equation in the form

$$q = A\Delta t/R$$

$R = x/k$ = 'thermal resistance' of the material.

By analogy with Ohm's law in electricity, a number of materials in series having the same cross-sectional area for heat transfer can be treated as resistances in series, and the following results

$$q = A\Delta t_o/R_o$$

Δt_o = overall temperature difference; R_o = overall 'resistance',
and $R_o = R_1 + R_2 + R_3 + \ldots\ldots = x_o/k_o$
$= (x_1/k_1) + (x_2/k_2) + (x_3/k_3) + \ldots\ldots$

Example 2.1

A drying oven is constructed of 5.0 mm thick mild steel sheets and is insulated with 25 mm thick magnesia. Calculate the rate of heat loss through the walls if the oven temperature is 60°C and the room temperature is 20°C. Thermal conductivity of mild steel = 45 W/m K; thermal conductivity of magnesia = 0.06 W/m K.

Thermal resistance of steel wall $R_1 = 0.005/45 = 0.000111$

Thermal resistance of magnesia $R_2 = 0.025/0.06 = 0.4167$

Overall resistance $R_o = R_1 + R_2 = 0.000111 + 0.4167$
$= 0.416811$

The rate of heat transfer through the two materials is given by

$q = A\Delta t_o/R_o = (60 - 20)/0.4168 = 96$ W/m^2 for 1 m^2 area

Heat Transfer Area

For a block of composite materials, the cross-sectional area for heat transfer will be the same for all sections of the block, thus

$q = A\Delta t_o/\Sigma(x/k)$

Thick wall tubes. For heat transfer through a thick wall tube, the area for heat transfer in a radial direction constantly changes along the radius, and thus an average area must be taken. It can be shown that the correct area to use is the log mean area (A_{lm})

$A_{lm} = (A_o - A_i)/\ln(A_o/A_i)$

A_o = outside surface area of tube; A_i = inside surface area of tube.

Thin wall tubes. For thin wall tubes, the surface area of the outside differs only slightly from the surface area of the inside, and an arithmetic mean area can be taken (A_m)

$A_m = \frac{1}{2}(A_o + A_i)$

Example 2.2

A steel pipe 25 mm bore x 3.0 mm thick contains condensing steam at 110°C. The rate of heat transfer through the surrounding layer of stagnant air is found to be 20 W/m^2K. If the room temperature is 15°C, calculate the heat loss per metre run of pipe. Thermal conductivity of steel = 45 W/m K.

Thermal resistance of the pipe $R_1 = 0.003/45 = 6.7 \times 10^{-5}$

Outside area of pipe $A_o = \pi \times 0.031 \times 1.0 = 0.09739\ m^2/m$

Inside area of pipe $A_i = \pi \times 0.025 \times 1.0 = 0.07854\ m^2/m$

Log mean area $A_{lm} = (0.09739 - 0.07854)/\ln(0.09739/0.07854)$

$= 0.08763\ m^2/m$ run.

The outside area of the pipe must be used in conjunction with the heat transfer coefficient of the stagnant air layer, and the overall rate of heat transfer is given by

$$q = \Delta t_o/(R_1/A_{lm} + 1/20A_o)$$
$$= (110 - 15)/(6.7 \times 10^{-5}/0.08763 + 1/20 \times 0.09739)$$
$$= 185 \text{ W/m run of pipe.}$$

If the arithmetic mean area of the pipe had been used in example 2.2, a value of 0.08797 m^2/m would have been obtained, giving an error for the overall rate of heat transfer of 0.0006%. For most thin wall pipes, the difference between the two means is negligible, but it is good practice to ensure that the error is small before using the arithmetic mean.

Example 2.3

If the steel pipe in example 2.2 were to be insulated with 35 mm thick magnesia, and the stagnant air heat transfer coefficient remained the same, what would be the new heat loss ? Thermal conductivity of magnesia = 0.06 W/m K.

The thermal resistance of the pipe and its mean area will be the same as in example 2.2, i.e. $R_1 = 6.7 \times 10^{-5}$, $A_{lm} = 0.08763\ m^2/m$. The mean area of the magnesia is evaluated as follows

Inside area = 0.09739 m^2/m - outside area of pipe

Outside area = $\pi \times 0.101 \times 1.0 = 0.3173\ m^2/m$

Log mean area $A'_{lm} = (0.3173 - 0.09739)/\ln(0.3173/0.09739)$

$= 0.1862\ m^2/m$.

Thermal resistance of magnesia $R_2 = 0.035/0.06 = 0.5833$

The outside area of the magnesia must now be used with the stagnant air heat transfer coefficient, and the overall rate of heat transfer will be

$$q = (110 - 15)/(6.7 \times 10^{-5}/0.08763 + 0.5833/0.1862 + 1/20 \times 0.3173) = 28.9 \text{ W/m run of pipe.}$$

If the arithmetic mean area had been used in example 2.3, the error introduced into the rate of heat transfer would have been 10.7% due to the magnesia being a thick wall tube. Ignoring the resistance of the pipe would have had a negligible effect on the calculated rate of heat transfer.

CONVECTION

Convection is the transfer of heat from one part of a fluid to another part of the same fluid, or from a solid surface to the bulk of a fluid in contact with the surface, by means of substantial molecular movement. Two modes of convection are recognised:

(a) natural convection, where the molecular movement is due to density differences generated by temperature gradients

(b) forced convection, where the motion is produced by external mechanical means, i.e. by pumping.

For practical heat transfer applications, the rates of heat transfer using forced convection are much better than for natural convection, and it is sound processing practice to always work using forced convection wherever possible.

Forced Convection Heat Transfer

If two fluids are flowing on either side of a solid wall, a plot of the temperature profiles on each side of the wall at steady-state conditions will appear similar to those shown in figure 2.1.

The temperature in the bulk of the fluid changes approximately parabolically near to the solid surface, and the distance over which this change takes place is known as the 'boundary' layer.

If we assume that all the resistance to heat transfer takes place over the boundary layers, i.e. that the temperature gradient

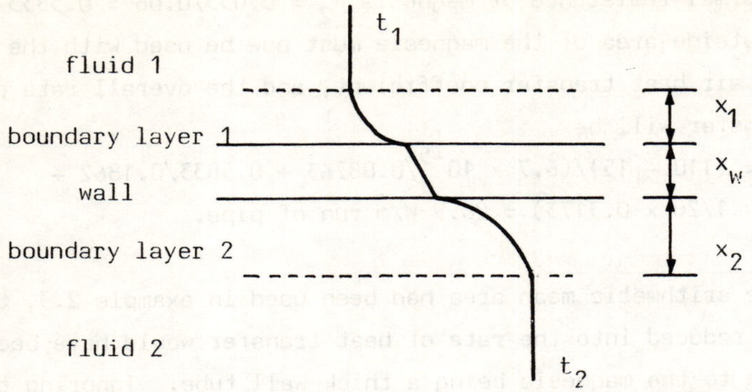

Figure 2.1 - Plot of temperature profiles near wall.

is a straight line, we can treat the problem as one of conduction through a composite layer of material, and the rate of heat transfer becomes

$$q = \Delta t_o/(x_1/k_1A_1 + x_w/k_wA_w + x_2/k_2A_2)$$

x = thickness of boundary layer; k = thermal conductivity of fluid; A = area for heat transfer; $\Delta t_o = t_1 - t_2$; subscript w refers to the solid wall.

The heat transfer coefficient (h) is defined as (k/x), and

h_1 = 'film heat transfer coefficient' for fluid 1 = k_1/x_1

h_2 = 'film heat transfer coefficient' for fluid 2 = k_2/x_2

and $q = \Delta t_o/(1/h_1A_1 + 1/h_2A_2 + x_w/k_wA_w) = \Delta t_o/\Sigma(1/hA)$

The thickness of the boundary layer varies depending on the rate of flow of fluid past the surface and the physical properties of the fluid. The thickness of the wall is known, and it is not necessary to use a film coefficient. Because the thickness of the boundary layers is small, there is little difference in areas A_1, A_2 and A_w, and it is common practice to use either the outside area or inside area of the wall for a thin wall tube and to be consistent in its use. Since one value for the area is used, the rate of heat transfer can be written as

$q = UA\Delta t_o$ where U = 'overall heat transfer coefficient'

and $1/U = 1/h_1 + 1/h_2 + x_w/k_w$

Dimensional analysis for forced convection heat transfer coefficients in pipe flow shows that the Nusselt number (Nu) is some function of the Reynolds number (Re) and the Prandtl number (Pr).

$Nu = hD/k$; $Re = Du\rho/\mu$; $Pr = C\rho\mu/k$

h = film heat transfer coefficient; D = pipe diameter; k = thermal conductivity of fluid; μ = fluid viscosity; Cp = fluid specific heat at constant pressure; ρ = fluid density.

A number of correlations are available in the literature based on this type of expression.

Laminar (streamline) flow inside pipes. For this flow region (Re < 2100), the following empirical relationship is commonly used [1]

$$Nu = 1.62(Re.Pr.D/L)^{1/3}$$

L = length of pipe; D = inside diameter of pipe. This relationship is valid for the following conditions

(a) Re.Pr.D/L > 120

(b) all physical properties must be evaluated at the mean bulk temperature (t_m) of the fluid. $t_m = \frac{1}{2}(t_o + t_i)$, and t_o = outlet temperature of fluid; t_i = inlet temperature of fluid.

Sieder and Tate [2] suggested that where large variations of viscosity with temperature exist (viscous oils etc.), the above expression is better modified to include a viscosity ratio

$$Nu = 1.86(Re.Pr.D/L)^{1/3}(\mu/\mu_s)^{0.14}$$

μ = viscosity at mean bulk temperature; μ_s = viscosity at wall temperature.

Turbulent flow in pipes. For this flow region, the expression due to Dittus and Boelter [3] is commonly used

$$Nu = 0.023(Re)^{0.8}(Pr)^n$$

n = 0.4 for heating; n = 0.3 for cooling. This relationship is only valid for

(a) Re > 10 000

(b) all physical properties are evaluated at the mean bulk temperature.

Colburn [4] proposed the following

$$St = 0.023(Re)^{-0.2}(Pr)^{1/3} \quad St = \text{Stanton number} = Nu/Re.Pr.$$

Colburn's expression can be transformed to

$$Nu = 0.023(Re)^{0.8}(Pr)^{1/3}$$

and this relationship is valid provided that

(a) the viscosity is evaluated at the mean film temperature, i.e. $\frac{1}{2}(t_s + t_m)$, where t_s = heating surface temperature

(b) all other properties are evaluated at t_m

(c) Re > 10 000

Sieder and Tate also suggested a modification to Colburn's expression for viscous liquids

$$Nu = 0.027(Re)^{0.8}(Pr)^{1/3}(\mu/\mu_s)^{0.14}$$

All of the above correlations apply to the flow of fluids inside pipes or tubes. For sections other than tubes, it has been found tha the use of the equivalent diameter (D_e) in evaluating the Reynolds number and Nusselt number gives reproducible results.

It should be noted that the accuracy of all heat transfer correlations is of the order of ± 15%, but in some cases an error of 50% can occur, particularly at the limits of validity.

Calculation of the rate of heat transfer in a heat exchanger involves the calculation of the overall heat transfer coefficient (U) using the correlations for the determination of the individual film heat transfer coefficients (h_1 and h_2).

Example 2.4

Water is heated using a jacketed pipe 25 mm in diameter from 20°C to 40°C. The jacket contains condensing steam at 95°C. Calculate the heat transfer coefficient for a water flowrate of 1.5 m^3/h. Density of water at 30°C = 998 kg/m^3; specific heat = 3.36 kJ/kg K; average thermal conductivity = 1.05 W/m K; viscosity at 30°C = 0.8 cP; viscosity at 62.5°C = 0.45 cP; viscosity at 95°C = 0.299 cP.

Using the Colburn correlation, the viscosity must be evaluated at the mean film temperature, all other properties at the mean bulk temperature.

Mean film temperature = $\{95 + \frac{1}{2}(40 + 20)\}/2 = 62.5$°C

Mean bulk temperature = $\frac{1}{2}(40 + 20) = 30$°C

Reynolds number $Re = (Du\rho/\mu)$; Prandtl number $Pr = (Cp\mu/k)$

Water velocity $u = (1.5 \times 4)/(3600 \times \pi \times 0.025^2) = 0.849$ m/s

$Re = (0.025 \times 0.849 \times 998)/(0.45 \times 10^{-3}) = 47072$

$Pr = (3.36 \times 10^3 \times 0.45 \times 10^{-3}/1.05) = 1.44$

$Nu = 0.023(Re)^{0.8}(Pr)^{1/3}$

and $h = 0.023(1.05/0.025)(47072)^{0.8}(1.44)^{1/3} = 5970$ W/m^2K

Using the Sieder and Tate correlation, the viscosity must be taken at the mean bulk temperature for Re and Pr.

$$Re = (0.025 \times 0.849 \times 998)/0.8 \times 10^{-3}) = 26478$$

$$Pr = (3.36 \times 10^{3} \times 0.8 \times 10^{-3}/1.05) = 2.56$$

$$Nu = 0.027(Re)^{0.8}(Pr)^{1/3}(\mu/\mu_s)^{0.14}$$

and $h = 0.027 \times 42\ (26478)^{0.8}(2.56)^{1/3}(0.8/0.299)^{0.14}$

$= 6150\ W/m^2K.$

The two values agree to within 3%, which is well within the accepted accuracy of the correlations.

Example 2.5

The jacketed pipe in example 2.4 has a steam side coefficient of 8000 W/m^2K. If the thickness of the tube is 3.0 mm, calculate the overall heat transfer coefficient. Thermal conductivity of the pipe material = 45 W/m K.

From example 2.4, $h_1 = 5970\ W/m^2K$ - Colburn's correlation.

$h_2 = 8000\ W/m^2K$

Overall heat transfer coefficient $U = 1/h_1 + 1/h_2 + x_w/k_w$

$$1/U = 1/5970 + 1/8000 + 0.003/45$$

$$= 0.0001675 + 0.000125 + 0.0000667 = 0.0003592$$

and $U = 2780\ W/m^2K.$

Agitated vessels. A number of correlations have been suggested for the case of heat transfer in agitated vessels using either jackets or internal coils. Chilton, Drew and Jebens [5] and Cummings and West [6] suggested the following

(a) for jacketed vessels

$$(h_iD/k) = 0.36(L^2N\rho/\mu)^{2/3}(Pr)^{1/3}(\mu/\mu_s)^{0.14}$$

h_i = inside surface heat transfer coefficient; D = inside diameter of the vessel; L = sweep diameter of the agitator; N = agitator speed of revolution, rev/unit time.

(b) for vessels containing internal coils, the combined results of Chilton *et al* and Cummings and West give

$$(h_oD/k) = 0.90(L^2N\rho/\mu)^{2/3}(Pr)^{1/3}(\mu/\mu_s)^{0.14}$$

h_o = heat transfer coefficient at the outside surface of the coil; D = inside diameter of the vessel.

Mean Overall Temperature Difference

Analysis of the heat exchanged by convection between two fluids separated by a solid wall shows that the correct overall temperature difference to use for both co- and counter-current flow, is the log mean temperature difference (Δt_{lm}).

$$\Delta t_{lm} = (\Delta t_1 - \Delta t_2)/\ln (\Delta t_1/\Delta t_2)$$

Δt_1 and Δt_2 = terminal temperature differences. Figure 2.2 represents a simple jacketed pipe heat exchanger operating in both co- and counter-current modes, and shows typical temperature profiles.

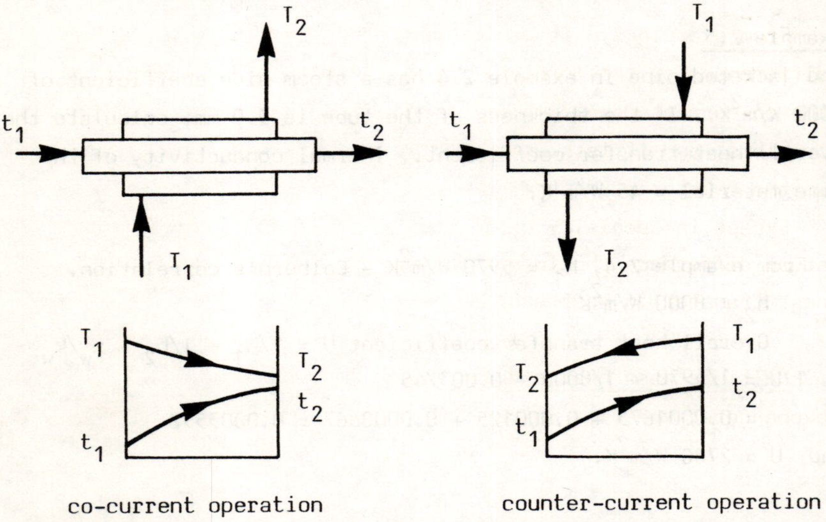

Figure 2.2 - Co- and counter-current exchanger operation.

Example 2.6

It is proposed to heat water from 20°C to 40°C using hot oil initially at 80°C. If the allowable outlet temperature of the oil is 50°C, what will be the mean overall temperature difference for a jacketed pipe heat exchanger operated (a) co-currently (b) counter-currently ?

(a) co-current case.

	End 1 Temperatures	End 2 Temperatures
Hot oil	80	50
Water	20	40

End 1 temperature difference $\Delta t_1 = 60°$

End 2 temperature difference $\Delta t_2 = 10°$

$\Delta t_{lm} = (60 - 10)/\ln(60/10) = 50/\ln(6) = 27.91°$

(b) counter-current case.

	End 1 Temperatures	End 2 Temperatures
Hot oil	80	50
Water	40	20

$\Delta t_1 = 40°$, $\Delta t_2 = 30°$

$\Delta t_{lm} = (40 - 30)/\ln(40/30) = 34.76°$

From example 2.6 it can be seen that the log mean temperature difference for the counter-current case is 24.5% higher than for the co-current case. This is generally true for all heat exchange problems, and it is sound practice to operate all heat exchangers in the counter-current mode whenever possible.

Except in specially defined cases, the arithmetic mean temperature difference should not be used for calculating the rate of heat transfer in heat exchangers.

Multiple pass heat exchangers. In the case of multiple pass heat exchangers, the flow is both co- and counter-current, and to take

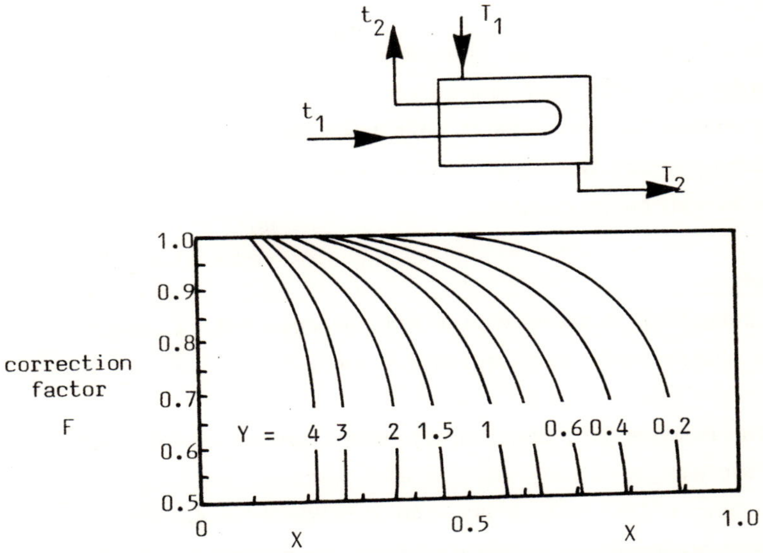

Figure 2.3 – Correction factor chart for a 2-tube pass heat exchanger.

this into account in design problems, Bowman, Mueller and Nagle [7] published charts of correction factors for a variety of configurations. These charts can be found in many textbooks, and are similar to figure 2.3.

The temperature difference for a mixed flow heat exchanger is calculated as if it were a true counter-current, single pass exchanger, and the temperature difference is then multiplied by the correction factor.

Example 2.7

A water stream is to be heated using hot oil at 100°C in a 2-tube pass heat exchanger. The inlet temperature of the water is 20°C, outlet temperature 80°C, and the hot oil leaves the heat exchanger at 90°C. Calculate the corrected log mean temperature difference.

For true counter-current flow

$\Delta t_{lm} = \{(90 - 20) - (100 - 80)\}/\ln(70/20) = 39.91°$

$X = (80 - 20)/(100 - 20) = 0.75 \quad = (t_2-t_1)/(T_1-t_1)$

$Y = (100 - 90)/(80 - 20) = 0.167 \quad = (T_1-T_2)/(t_2-t_1)$

From figure 2.3, the correction factor F = 0.93

Corrected $\Delta t_{lm} = 0.93 \times 39.91 = 37.11°$

Having calculated the individual film heat transfer coefficients, the overall coefficient and corrected log mean temperature difference, either (a) the rate of heat transfer from a given heat exchanger area, or (b) the area required for a specified rate of heat transfer can be calculated.

Example 2.8

What length of 25 mm bore pipe will be required to perform the water heating duty in examples 2.4 and 2.5 ?

From example 2.5, the overall heat transfer coefficient U = 2784 W/m^2K.

$\Delta t_{lm} = \{(95 - 20) - (95 - 40)\}/\ln(75/55) = 64.48°$

Using a single pass jacketed tube, no correction to Δt_{lm} is required.

$$q = UA\Delta t_{lm} = WC_p(t_o - t_i)$$

$$2784A \times 64.48 = (1.5 \times 998 \times 3.36 \times 10^3[40 - 20])/3600$$

$$= 27944 \text{ W}$$

thus $A = 0.1557 \text{ m}^2 = \pi DL$

and $L = 0.1557/0.025\pi = 1.98$ m.

2.2 UNSTEADY-STATE HEAT TRANSFER

CONDUCTION

A substance changes in temperature when the amount of heat flowing into it is different from the heat flowing out of it (assuming there are no other thermal energy sources within the substance, such as those associated with chemical reactions or phase changes). Because there is a continuously changing temperature and heat flux at all points within the substance this situation is described as unsteady-state heat transfer.

In a solid material within which heat is transferred solely by conduction, heat conducted into and out of an infinitesimally small element of the material can be equated to the rate of change of temperature of that element. From this it can be shown that

$$\frac{\partial t}{\partial \theta} = \frac{k}{c\rho}\left[\frac{\partial^2 t}{\partial x^2} + \frac{\partial^2 t}{\partial y^2} + \frac{\partial^2 t}{\partial z^2}\right]$$

i.e. (rate of change of temperature with time)= $\frac{k}{c\rho}$ (sum of rates of rate of change of temperature with position in the x, y and z directions).

The term $k/c\rho$ is called the thermal diffusivity (a) and it has dimensions of m^2/s. If the heat flow is only in one direction (x) the above equation becomes $\partial t/\partial\theta = a(\partial^2 t/\partial x^2)$. The solution of this differential equation to give the temperature at any time and position within a solid material undergoing heating or cooling requires a definition of the relevant boundary conditions. For example, consider the case of a slab of material of initially uniform temperature t_o. The two largest parallel faces of the slab are brought instantaneously to a temperature t_1, causing heat to flow in a direction perpendicular to these faces. If heat flow in any other direction is ignored the solution of the differential equation gives:

$$\frac{t-t_1}{t_o-t_1} = \sum_{n=o}^{\infty} \frac{4}{(2n+1)\pi} \exp\left[\frac{-(2n+1)^2\pi^2 a\Theta}{L^2}\right] \sin\left[\frac{(2n+1)\pi x}{L}\right]$$

where t is the temperature at a point in the slab distance x from the surface at a time Θ, n is an integer and L is the thickness of the slab.

The assumptions made in the derivation of this equation are (a) the material is homogeneous and of constant physical properties, (b) the surface temperature t_1 is achieved instantaneously and is invariate and (c) there is no surface thermal resistance i.e. an infinite heat transfer coefficient at the surface. Figure 2.4 shows this solution in graphical form. Figure 2.5 shows the same solution in the form of a heating (or cooling) curve for four positions in the material.

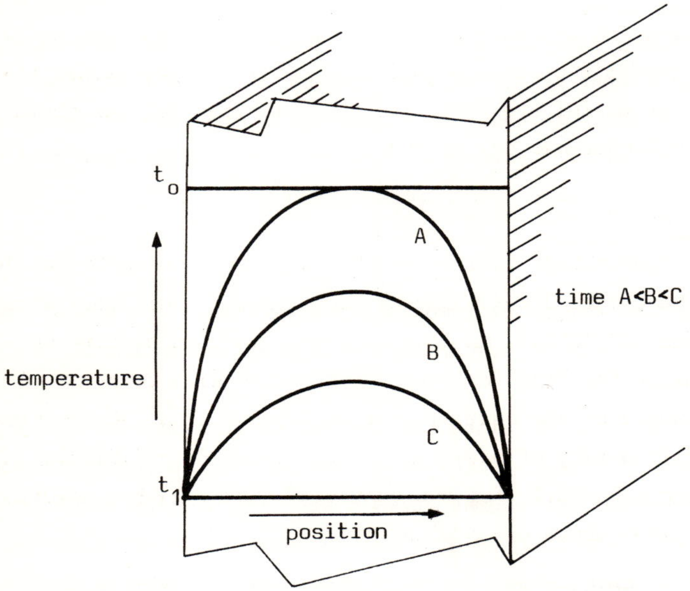

Figure 2.4 Temperature profiles in an infinite slab.

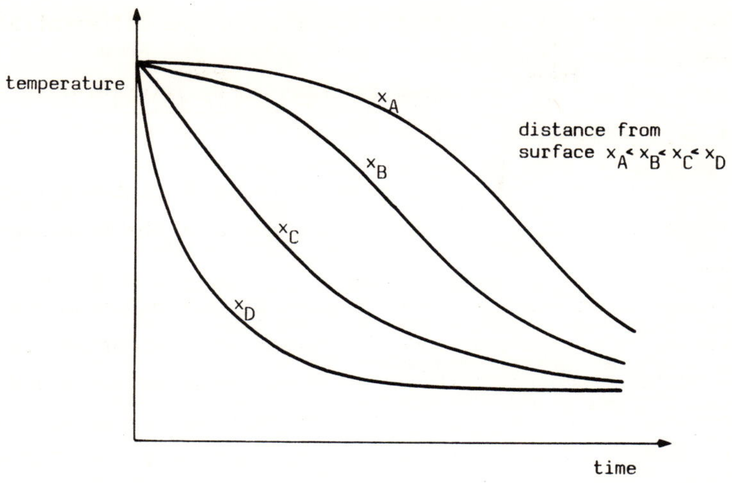

Figure 2.5 Temperature change with time in heating (cooling) an infinite slab.

Analytical solutions of the differential equation for other boundary conditions are available (e.g. Carslaw and Jaeger [8]). Particularly useful are those for the infinite slab (as above), the infinite cylinder and the sphere.

Example 2.9

A slab of frozen beef (thickness 80 mm, thermal conductivity 1W/mK, specific heat 1600 J/kgK, density 800 kg/m^3) is initially at a temperature of -10°C. The slab is placed in a cold store at -30°C.

(a) Calculate the time for the centre of the slab to reach -25°C.

(b) What would be the equivalent time for a slab of 50 mm thickness?

Thermal diffusivity of material (a) = 1/1600x800 = $7.813\times10^{-7} m^2/s$

(a) Putting x/L = 0.5 (for the centre of the slab) the solution for the slab given above expands to:

$$\frac{t-t_1}{t_o-t_1} = \frac{4}{\pi}\exp\left(\frac{-\pi^2 a\theta}{L^2}\right) - \frac{4}{3\pi}\exp\left(\frac{-9\pi^2 a\theta}{L^2}\right) + \frac{4}{5\pi}\exp\left(\frac{-25\pi^2 a\theta}{L^2}\right) - \ldots$$

Substituting t_o = -10°C, t_1 = -30°C, a = $7.813\times10^{-7} m^2/s$ and L = 0.08m:

$$(t+30)/20 = 1.273\exp(-1.205\times10^{-3}\theta) - 0.4244\exp(-1.084\times10^{-2}\theta) + 0.2546\exp(-3.012\times10^{-2}\theta) - \ldots$$

Therefore $t = [25.46\exp(-1.205\times10^{-3}\theta) - 8.488\exp(-1.084\times10^{-2}\theta) + \ldots] - 30.$

To find θ when $t = -25°C$ a suitable method is to substitute a number of values of θ in the equation to give values of t around -25. The value of θ for $t = -25°C$ can then be obtained by interpolation from a graph of t against θ. In this case $\theta = 1000$(s) gives $t = -22.37$; $\theta = 1500$ gives $t = -25.82$ and $\theta = 2000$ gives $t = -27.71$. By interpolation $\theta = 1350$ for $t = -25$. In calculating t it will be seen that only the first term of the series solution is significant. So in this case it is possible to write $t = 25.46 \exp(-1.205 \times 10^{-3}\theta) - 30$ and this gives the required time of 1350s (22.5 mins) directly.

(b) For the same centre temperature to be achieved in a slab of different thickness all the exponential terms must remain the same; this can only occur if θ/L^2 remains constant, i.e. time $\propto$ (thickness)^2 Therefore, for a slab of 50 mm thickness the time for a centre temperature of -25°C will be $1350 \times (50^2/80^2) = 530$s.

Note that in this example only the first term of the series solution was significant. This is often the case in real systems - it is normally necessary to include a larger number of terms only for temperatures close to the initial value.

Graphical solutions and surface thermal resistance.

The complexity of the unsteady-state heat transfer equations makes them tedious to use. An alternative to direct application of the equations is to formulate them graphically. A number of these graphs have been produced and figures 2.6 - 2.8 show their typical form. These graphs have universal applicability for all materials of the defined geometries because they are expressed in dimensionless form, i.e. instead of directly plotting the time/temperature/position relationship as in figures 2.4 and 2.5 in this case 'time' is expressed as $a\theta/L^2$(the Fourier Number), 'temperature' as $(t-t_1)/(t_0-t_1)$ and 'position' as x/L. (In these figures the dimension L is replaced by the half-thickness of the slab(δ), the radius of the cylinder (R) and the radius of the sphere (S).)

Another advantage of these graphical solutions is that they include the effect of convective heat transfer at the surface of the solid material. This is probably the most important assumption, in practical terms, of the derivation referred to earlier. The heat transfer coefficient is allowed for in the term m in the graphs,

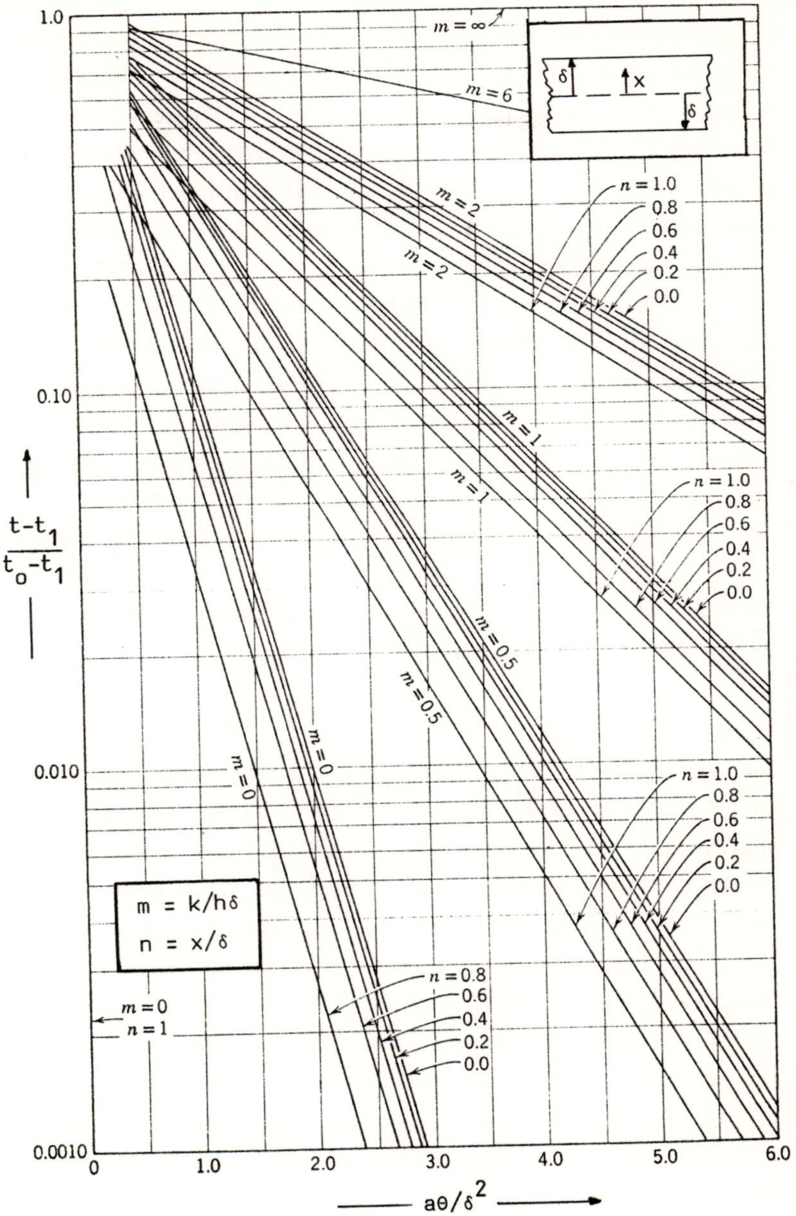

Figure 2.6 Conduction in an infinite slab (from [32])

(Values for $(t-t_1)/(t_o-t_1)>0.3$ are given in figure 2.6a)

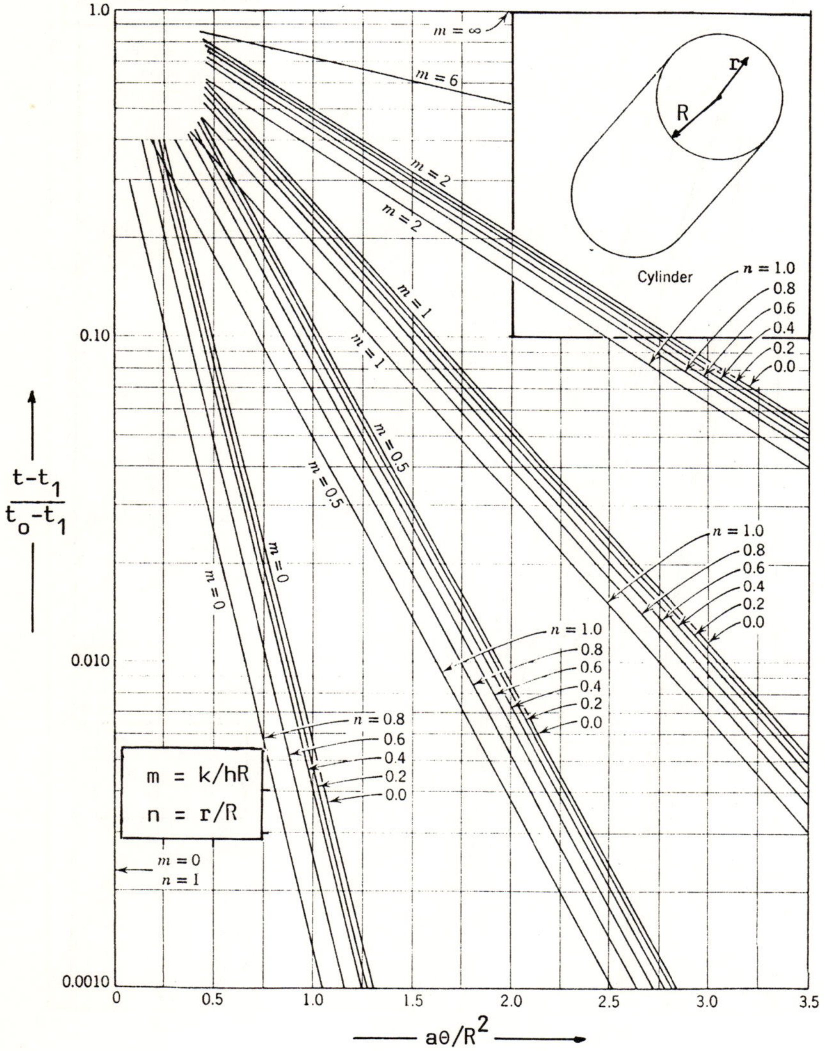

Figure 2.7 Conduction in an infinite cylinder (from [32])

(Values for $(t-t_1)/(t_o-t_1)>0.3$ are given in figure 2.7a)

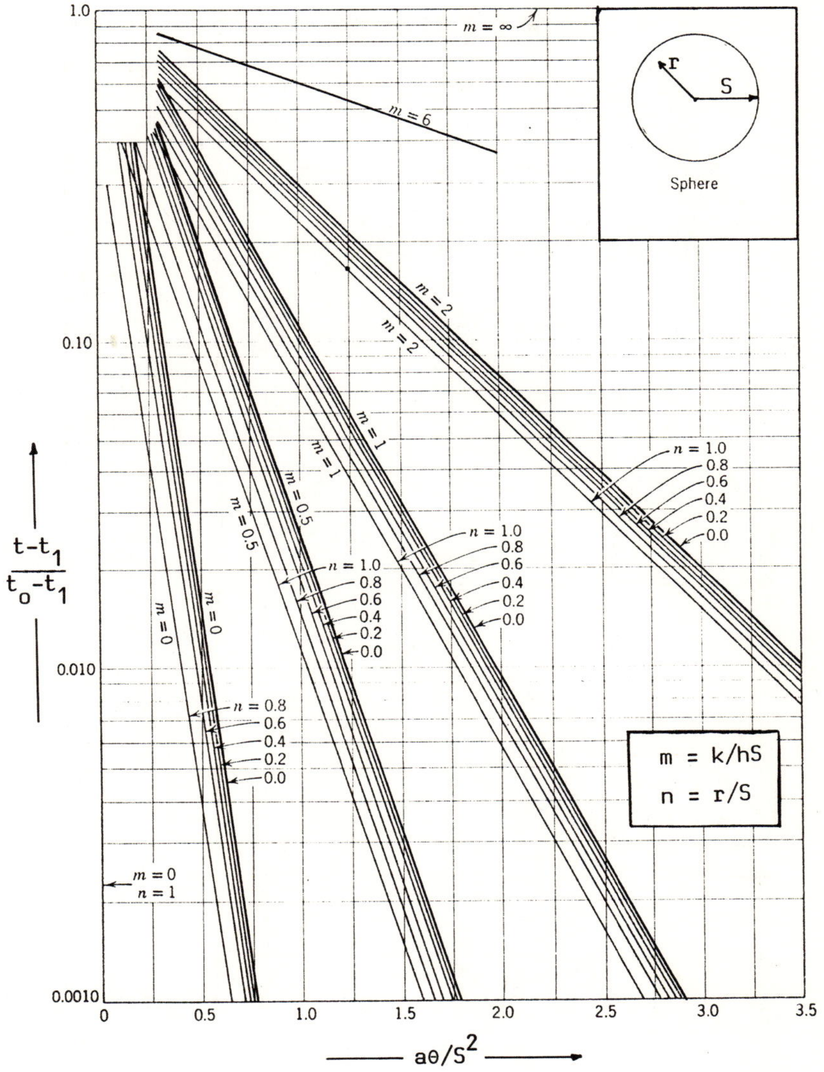

Figure 2.8 Conduction in a sphere (from [32])

(Values for $(t-t_1)/(t_o-t_1)>0.3$ are given in figure 2.8a)

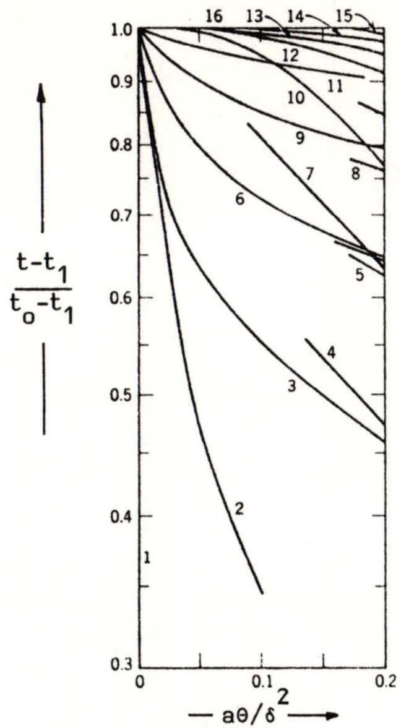

Figure 2.6a
INFINITE SLAB

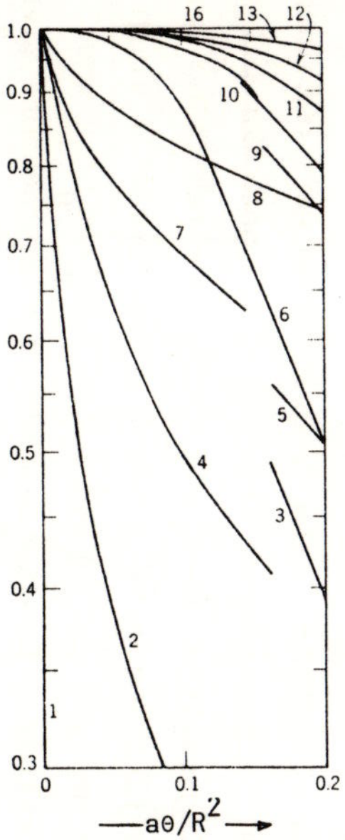

Figure 2.7a
INFINITE CYLINDER

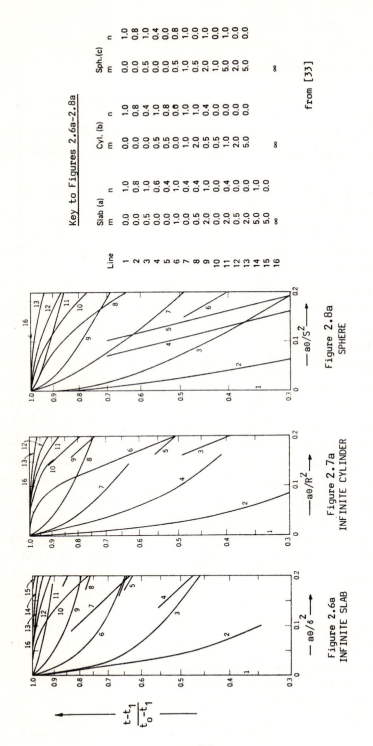

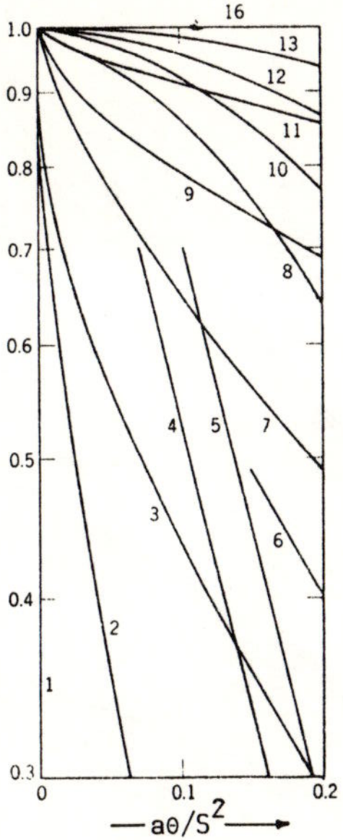

Figure 2.8a
SPHERE

Key to Figures 2.6a-2.8a

	Slab (a)		Cyl. (b)		Sph.(c)	
Line	m	n	m	n	m	n
1	0.0	1.0	0.0	1.0	0.0	1.0
2	0.0	0.8	0.0	0.8	0.0	0.8
3	0.5	1.0	0.0	0.4	0.5	1.0
4	0.0	0.6	0.5	1.0	0.0	0.4
5	0.0	0.4	0.5	0.8	0.0	0.0
6	1.0	1.0	0.0	0.0	0.5	0.8
7	0.0	0.4	1.0	1.0	1.0	1.0
8	0.5	0.4	2.0	1.0	0.5	0.0
9	2.0	1.0	0.5	0.4	2.0	1.0
10	0.0	0.0	0.5	0.0	1.0	0.0
11	2.0	0.4	1.0	0.0	5.0	1.0
12	0.5	0.0	2.0	0.0	2.0	0.0
13	2.0	0.0	5.0	0.0	5.0	0.0
14	5.0	1.0				
15	5.0	0.0				
16	∞		∞		∞	

from [33]

defined as k/hL, i.e. a dimensionless ratio of the relative magnitude of conductive and convective effects.

Example 2.10

Recalculate the problem of example 2.9 using the graphical solutions, and assuming heat transfer coefficients of 40, 400 and 4000 W/m^2K.

Using figure 2.6 for an infinite slab and δ = half-thickness of slab = 0.04m:

$n = x/\delta = 0$ (for centre), $(t-t_1)/(t_o-t_1) = 5/20 = 0.25$.

For $h = 40W/m^2K$, $m = 0.625$ and $a\theta/\delta^2 = 1.5$ (by interpolation from graph between m = 0.5 and m = 1)

Hence $\theta = 1.5 \times 0.04^2/7.813 \times 10^{-7} = 3070s$

For $h = 400\ W/m^2K$, $m = 0.0625$ and $a\theta/\delta^2 = 0.8$ (by interpolation between m = 0 and m = 0.5)

Hence $\theta = 1640s$

For $h = 4000\ W/m^2K$, $m = 0.00625$ and $a\theta/\delta^2 = 0.7$ (taking $m \doteq 0$)

Hence $\theta = 1430s$

This last figure should be identical to that obtained in example 2.9 since h is sufficiently large to be considered equivalent to a negligible surface resistance. The figures given were obtained from simple visual interpolations from the graph both for values of m and on the log scale of $(t-t_1)/(t_o-t_1)$. The resulting errors can be reduced by plotting $a\theta/\delta^2$ against the values of m given on the graph for the desired single value of $(t-t_1)/(t_o-t_1)$ and interpolating from this. In this example such a procedure gives times of 3120s, 1570s and 1390s for h = 40, 400 and 4000 W/m^2K respectively.

Finite shape bodies.

It is possible to calculate the temperature within finite cylinders and brick-shaped bodies by use of the following procedure:

for a finite cylinder of diameter D, height H -

$$\frac{t-t_1}{t_o-t_1} = \left(\frac{t-t_1}{t_o-t_1}\right)_{\text{infinite cylinder, diameter D}} \times \left(\frac{t-t_1}{t_o-t_1}\right)_{\text{infinite slab, thickness H}}$$

for a brick-shaped body of major lengths L_1 and L_2, and ignoring heat flow through the 'edges' -

$$\frac{t-t_1}{t_o-t_1} = \left(\frac{t-t_1}{t_o-t_1}\right)_{\substack{\text{infinite slab,}\\ \text{thickness } L_1}} \times \left(\frac{t-t_1}{t_o-t_1}\right)_{\substack{\text{infinite slab,}\\ \text{thickness } L_2}}$$

Example 2.11

A can of food (the mode of heat transfer within which is solely conduction) is sterilised in steam at 130°C. The can dimensions are 10 cm diameter, 15 cm height. The can contents have a thermal conductivity of 0.48 W/mK, specific heat of 3000 J/kg K, density 800 kg/m^3 and an initial temperature of 30°C. Calculate the temperature after 90 minutes (a) at the centre of the can, (b) at a point 3.5 cm from the centre and 5 cm from an end, (c) at the centre of the can if the heating were carried out in a medium giving a heat transfer coefficient of 20 W/m^2K.

$a = 0.48/3000 \times 800 = 2 \times 10^{-7} m^2/s$

(a) Using figure 2.7 for an infinite cylinder, R = radius = 0.05m:
$n = r/R = 0$ (at can centre); $a\theta/R^2 = 2\times10^{-7} \times 5400/0.05^2 = 0.432$
Therefore $(t-t_1)/t_o-t_1)_{\text{inf.cylinder}} = 0.141$
Using figure 2.6a for an infinite slab, δ = half-thickness = 0.075m
$n = 0$ (at can centre); $a\theta/\delta^2 = 2\times10^{-7} \times 5400/0.075^2 = 0.192$
Therefore $(t-t_1)/(t_o-t_1)_{\text{inf. slab}} = 0.78$
Hence $(t-t_1)/(t_o-t_1)_{\text{finite cylinder}} = 0.141 \times 0.78 = 0.110$
and $t = 130 - (0.110 \times 100) = 119$°C

(b) Using figure 2.7 : $n = 3.5/5 = 0.7$; $a\theta/R^2 = 0.432$ (as before),
therefore $(t-t_1)/(t_o-t_1)_{\text{inf. cylinder}} = 0.0518$
Using figure 2.6a: $n = x/\delta = 2.5/7.5 = 0.33$; $a\theta/\delta^2 = 0.192$ (as before)
therefore $(t-t_1)/(t_o-t_1)_{\text{inf. slab}} = 0.68$
Hence $(t-t_1)/(t_o-t_1)_{\text{finite cylinder}} = 0.0518 \times 0.68 = 0.0352$
and $t = 130 - (0.0352 \times 100) = 126.5$°C

(c) In this case $m = k/hR = 0.48/(20\times0.05) = 0.48$ for an infinite cylinder
and $m = k/h\delta = 0.48/(20\times0.075) = 0.32$ for an infinite slab

Using the same procedure as in (a), $(t-t_1)/(t_o-t_1)_{inf.\ cylinder} = 0.46$

$(t-t_1)/(t_o-t_1)_{inf.slab} = 0.84$

Hence $(t-t_1)/t_o-t_1)_{finite\ cylinder} = 0.46 \times 0.84 = 0.386$

and $t = 130 - (0.386 \times 100) = 91.4^oC$.

It should be noted that when using the graphical solutions to give $(t-t_1)/(t_o-t_1)$ values above about 0.4, and particularly with low values of m, the graphs given here do not allow precise readings to be taken. The graphs given by Schneider [9] are recommended for more accurate evaluation.

CONVECTION

A common heat transfer situation occurs where a batch of liquid is heated or cooled by a surrounding heat transfer medium. Heat will be transferred to the liquid by convection and, since the rate of heat transfer depends on the temperature difference between the heat transfer medium and the liquid, this rate will be continuously changing as the liquid temperature changes. This is unsteady-state convection. If the initial temperature of the liquid is uniform at t_o and the heating medium temperature is t_1, it can be shown that

$$(t-t_1)/(t_o-t_1) = \exp(-UA\theta/mc)$$

where t is the liquid temperature at time θ, U is the overall heat transfer coefficient between the heating medium and the liquid, A is the heating surface area, m and c are the mass and specific heat of the liquid.

The equation can be conveniently rearranged to give

$$\theta = (mc/UA) \ln[(t_1-t_o)/(t_1-t)]$$

These equations assume that the liquid is well agitated so that there are no temperature gradients within it.

Example 2.12

A batch of milk, volume 250 litres, is heated from an initial temperature of 4°C in a steam-jacketed, agitated, vessel of heating surface area $1.5m^2$.

(a) Calculate the time taken to heat the milk to 70°C, if the overall heat transfer coefficient is 800 W/m^2K. The specific heat of the milk is 3900 J/kgK and its density is 1030 kg/m^3. The steam temperature is 130°C. (b) Suggest possible changes in the processing conditions to reduce the heating time by 25%.

(a) Heating time $=[(0.25\times1030\times3900)/(800\times1.5)]\ln[(130-4)/(130-70)]$

$= 621s = 10.3$ minutes

(b) (i) increase steam temperature:

$(t-t_1)/(t_o-t_1) = \exp[(-800\times1.5\times466)/(0.25\times1030\times3900)]$

$= 0.5730$ for $\theta = .75\times621 = 466s$

Therefore steam temperature (t_1) would have to be 159°C.

(ii) increase heat transfer coefficient:

since $\theta \propto 1/U$, the heat transfer coeffient would have to be increased to $800\times(4/3) = 1067 W/m^2K$. The most likely procedure for this increase would be to increase the agitator speed. From the correlation for jacketed vessels given in Section 2.1 it can be seen that the inside heat transfer coeffient is proportional to (agitator speed)$^{2/3}$. Therefore it would be necessary to increase the agitator speed to $(4/3)^{1.5}$ of its original value, i.e. by 54%. This assumes a negligible resistance to heat transfer on the steam side.

2.3 NON-NEWTONIAN HEAT TRANSFER

Since most non-Newtonian fluids flowing in a piping system exhibit high values for the 'apparent' viscosity, it is unusual to find that fully-developed, turbulent flow is achieved. As a result, the correlations which have been developed for heat transfer to such fluids are modifications of the correlations which have been developed for heat transfer to highly viscous, Newtonian fluids flowing under laminar conditions, i.e. of the form

$$h = \varphi(Re.Pr.D/L)$$

h = heat transfer coefficient; Re = Reynolds number ($Du\rho/\mu$); Pr = Prandtl number ($C_p\mu/k$).

It is possible to obtain rigid, mathematical solutions for the heat transfer coefficient for a non-Newtonian fluid by applying the complex flow equations to the heat transfer correlations, but these solutions lead to very elegant mathematical expressions, which, for design purposes, become extremely laborious to apply.

The term (Re.Pr.D/L) is very useful in the case of non-Newtonian fluids, since a numerical value for the 'viscosity' is not required, and this particular physical property is probably the most indeterminate of all for non-Newtonian fluids.

$$(Re.Pr.D/L) = (Du\rho/\mu)(c_p\mu/k)\ (D/L) = (D^2u\rho/kL)$$

This term is closely related to the Graetz number (Gz)

$$Gz = (Wc_p/kL) \quad \text{and} \quad (D^2u\rho) = 4W/\pi, \ W = \text{mass flowrate.}$$

Thus, the main correlations for non-Newtonian heat transfer resolve into some function of the Graetz number, with modifications depending on the type of fluid being treated, e.g. power law fluid, Bingham plastics etc.

Because of the approximate nature of the correlations, the mean heat transfer coefficient (h_m) is invariably used with the arithmetic mean temperature difference (Δt_m)

$$\Delta t_m = t_w - \tfrac{1}{2}(t_o + t_i)$$

t_w = heating surface temperature; t_o = mean bulk temperature of the fluid leaving the heated section; t_i = mean bulk temperature of the fluid entering the heated section.

Laminar Flow in Pipes with No Natural Convection

Pigford [10] suggests the use of the following equation for Bingham plastic, power law and Newtonian fluids

$$(h_mD/k) = 1.75\delta^{1/3}(Gz)^{1/3} \quad \text{and} \quad \delta = (3n' + 1)/4n'$$

n' = generalised rheological parameter.

For Bingham plastics, n' must be known at the working shear rate;

for power law fluids, n' = n = flow behaviour index;

for Newtonian fluids, n' = 1.

This equation is valid for Gz>20 and n'>0.10[11], and has been found to be accurate to ±15% for Gz>100, when natural convection effects become minimal [12].

Example 2.13

A power law fluid is flowing through a 25 mm bore copper pipe at a mass flowrate of 285 kg/h and a temperature of 38°C. After a sufficiently long section to ensure a fully-developed velocity profile, the fluid enters a 2.0 m long annular jacketed section containing condensing steam at 93°C. The fluid is characterised by the equation

$$K = 0.75(\exp E/RT)^n \text{ where } E/R = 4090 \text{ and } n = 0.4$$

K = fluid consistency; E = 'activation energy', J/mol; R = universal gas constant, J/mol Kelvin; T = absolute temperature, Kelvin. Mean fluid density in the heating section = 1200 kg/m^3; mean specific heat = 3.36 kJ/kg K; thermal conductivity = 1.20 W/m K. Determine (a) the mean heat transfer coefficient (b) the bulk temperature of the fluid leaving the heating section.

(a) Using Pigford's correlation

$$\delta = (3 \times 0.4 + 1)/(4 \times 0.4) = 1.375$$

$$Gz = (285 \times 3.36 \times 10^3)/(1.20 \times 2.0 \times 3600) = 110.8$$

Since the value of the Graetz number is greater than 100, the correlation is valid.

$$h_m = (1.75 \times 1.20 \times 1.375^{1/3} \times 110.8^{1/3})/0.025 = 448.6 \text{ W/m}^2\text{K.}$$

(b) A heat balance over the heating section will give

$$q = Wc_p(t_o - t_i) = h_m A \Delta t_m$$

thus $q = 448.6 \times \pi \times 0.025 \times 2.0[93 - \frac{1}{2}(t_o + 38)] = 70.48(74 - \frac{1}{2}t_o)$

and $q = 285 \times 3.36 \times 10^3(t_o - 38)/3600 = 266(t_o - 38)$

so $t_o(266 + 35.24) = (10108 + 5215.5)$

and the bulk mean outlet temperature of the fluid = 51° C.

Power law fluids. Metzner and Gluck [13] suggested for power law fluids

$$(h_m D/k) = 1.75\delta^{1/3}(Gz)^{1/3}(K_b/K_w)^{0.14}$$

K_b = fluid consistency at the bulk mean fluid temperature; K_w = fluid consistency at the heating surface temperature.

This correlation is valid for the range $50 < Gz < 5000$, and has been found to be accurate to ±12% [14]

Example 2.14

The Metzner and Gluck correlation will be used to solve the problem in example 2.13 above.

Since t_o is unknown, a value for K_b cannot be directly calculated, and an iterative procedure must be adopted. Since the ratio of the fluid consistencies is raised to the power 0.14, the simplest assumption for a first guess is that $K_b = K_w$.

First approximation.

$K_b/K_w = 1.0$ and $(K_b/K_w)^{0.14} = 1.0$

$\delta = 1.375$, Gz = 110.8 and $h_m = 448.6\ \mathrm{W/m^2K}$ (as example 2.13).

A heat balance will give the same result as example 2.13, and a first approximation for $t_{o1} = 50.9^{\circ}C$.

Second approximation.

We can now use the estimated mean bulk outlet temperature from above to give a value for K_{b1}.

At 93°C, T = 366 K; $K_w = 0.75(\exp 4090/366)^{0.4} = 0.75e^{4.47}$
$= 65.52$

Mean bulk temperature = ½(38 + 50.9) = 44.5°C = 317.5 K.

$K_{b1} = 0.75(\exp 4090/317.5)^{0.4} = 0.75e^{5.15} = 129.7$

$K_{b1}/K_w = 129.7/65.52 = 1.98$; $(K_{b1}/K_w)^{0.14} = 1.10$

Therefore, $h_{m2} = 1.10h_{m1} = 493.5\ \mathrm{W/m^2K}$. Using this value in a heat balance gives a value for $t_{o2} = 52.0°C$.

A third approximation gives

$h_{m3} = 493.0\ \mathrm{W/m^2K}$ and $t_{o3} = 52.0°C$.

Thus, (a) the mean heat transfer coefficient = 493.0 $\mathrm{W/m^2K}$.

(b) mean fluid outlet temperature = 52.0°C.

Comparison of the values obtained for the mean heat transfer coefficient between Pigford's correlation and that of Metzner and Gluck shows that within the stated confidence limits there is good agreement. However, care must be exercised in their use for design purposes by ensuring that Gz>100 to minimise natural convection.

Bingham plastic fluids. An approximation suggested by Hirai [15] is

$$(h_m D/k) = 1.75(Gz)^{1/3}[3(1 - c)/(c^4 - 4c + 3)]^{1/3}$$

c = ratio of yield stress to shear stress at the wall.

This approximation is valid for $Gz[3(1 - c)/(c^4 - 4c + 3)]>100$, and the accuracy of correlation has been found to be ±15%

However, to minimise natural convection effects, the Graetz number should be greater than 100.

Example 2.15

A Bingham plastic fluid is to be heated in a 25 mm bore jacketed pipe using 93°C steam. The mass flowrate of the fluid is 355 kg/h and the inlet temperature is 38°C. Estimate the length of heating section required to raise the temperature of the fluid by 12°, and calculate the mean heat transfer coefficient. Mean density = 1040 kg/m^3; mean thermal conductivity = 1.20 W/m K; mean specific heat = 3.36 kJ/kg K; yield stress at 44°C = 7.10 N/m^2; mean stress at wall = 15.0 N/m^2.

$c = 7.10/15.0 = 0.473$ and $3(1 - c)/(c^4 - 4c + 3) = 1.831$

$Gz = 355 \times 3.36 \times 10^3/(1.20 \times 3600L) = 276/L$ - the length of section being unknown.

Using Hirai's correlation

$$h_m = 1.75 \times 1.20(276/L \times 1.365)^{1/3}/0.025$$
$$= 84(376.7/L)^{1/3} = 669\ /(L)^{1/3}$$

An overall heat balance will give

$$Wc_p(12) = h_m A\Delta t_m = h_m \pi DL\Delta t_m$$

and $h_m L = (355 \times 3.36 \times 10^3 \times 12)/(3600 \times \pi \times 0.025[93 - 88/2])$

$$= 1033$$

so $h_m = 1033/L$

Combining both expressions for h_m

$$L^{2/3} = 1033/669 = 1.544$$
$$L = 1.92 \text{ m.}$$

The important item left outstanding is to check on the validity of the correlation

$$Gz = 144$$

thus, the correlation is valid, and natural convection effects will be negligible.

Hence, $h_m = 538$ W/m^2K and $L = 1.9$ m.

In the design of heat transfer equipment for the laminar flow of non-Newtonian fluids, the following points are of extreme importance

(1) the mean heat transfer coefficient (h_m) is a function of the

Graetz number, and is therefore a function of (1/L). Thus, long tube heat exchangers do not give the most effective heat transfer rates.

(2) by combining the correlation with a heat balance for the case of a jacketed pipe, the following expressions result

(a) Pigford's correlation with jacketed pipe.

$$Gz = [1.75\pi\Delta t_m/(t_o - t_i)]^{3/2}\sqrt{(\delta)}$$

(b) For the Metzner and Gluck correlation.

$$Gz = [1.75\pi\Delta t_m/(t_o - t_i)]^{3/2}(K_b/K_w)^{0.21}\sqrt{(\delta)}$$

(c) Hirai's approximation for Bingham plastic fluids.

$$Gz = [1.75\pi\Delta t_m/(t_o - t_i)]^{3/2}\sqrt{[3(1 - c)/(c^4 - 4c + 3)]}$$

It can be seen that, since the correlations are valid for Gz greater than 100, it is better to use a series of short jacketed pipes, rather than one long single pipe. Also, the maximum bulk outlet temperature for any given flowrate and inlet temperature will be achieved when the value of the Graetz number is approximately 100 for the heated length (in order to minimise natural convection effects).

Example 2.16

Taking example 2.13 as a basis, what will be the maximum outlet temperature theoretically possible for a single heated pipe, and what will be the length of pipe required to achieve this?

By combining Pigford's correlation with the heat balance

$$[100/\sqrt{(1.375)}]^{2/3} = 1.75\pi[93 - \tfrac{1}{2}(t_o + 38)]/(t_o - 38)$$

so $3.524(t_o - 38) = (74 - \tfrac{1}{2}t_o)$

$t_o = 51.7°C$, and L = 2.2 m.

In order to heat a fluid to a given temperature at a certain flowrate, a series of jacketed pipes operating at a value of Gz = 100 is usually the most effective method, provided that a sufficient length of unheated pipe is used to join the heated sections to allow the flow pattern to re-develop.

For Newtonian fluids in laminar flow, the entrance length (L_e) required for the flow pattern to fully develop is given by Schiller [16] as

$$L_e = 0.029D(Re).$$

Bogue [17] extended this expression for power law fluids

(a) for pseudoplastic fluids ($n<1$), L_e is shorter than for Newtonian fluids

(b) for dilatant fluids ($n>1$), L_e is correspondingly greater.

Example 2.17

It is required to heat the power law fluid in example 2.13 to 75°C. Assuming that the physical properties do not substantially change, suggest a suitable system.

A suitable system could consist of a series of jacketed pipes, all operating at a Graetz number of 100, and joined by unheated sections of pipe to allow re-development of the flow pattern.

First section.

From example 2.16, the maximum outlet temperature = 51.7°C and the length of heated section = 2.2 m.

Second section.

$$t_i = 51.7°C, \; Gz = 100.$$

Combining the Pigford correlation and a heat balance

$$3.524(t_o - 51.7) = [93 - \tfrac{1}{2}(t_o + 51.7)]$$

$$t_o = 61.9°C, \; L = 2.2 \text{ m}.$$

Third section.

$$t_i = 61.9°C, \; Gz = 100 \text{ and } t_o = 69.6°C, \; L = 2.2 \text{ m}.$$

Fourth section.

$$t_i = 69.6°C, \; Gz = 100 \text{ and } t_o = 75.4°C, \; L = 2.2 \text{ m}.$$

Thus, in order to heat the fluid from 38°C to 75°C, four 2.2 m long sections of jacketed 25 mm bore pipe will be required, each joined by an unheated section of pipe. If the value of the Reynolds number is 1000, the maximum unheated length of pipe required between each section would be 0.725 m.

2.4 PLATE HEAT EXCHANGER

The plate heat exchanger consists of a set of thin rectangular plates separated by gaskets so that narrow channels are formed through which fluids can be pumped. There are ports at the corners of the plates which determine the flow path of the fluids. Figure 2.9 shows one arrangement of flow conditions between the plates.

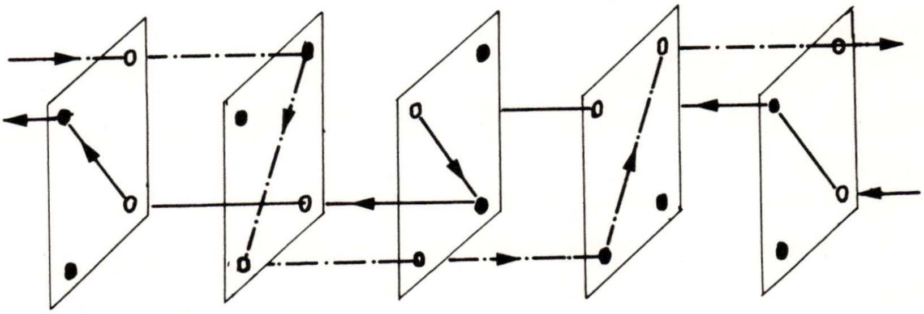

Figure 2.9 - Flow paths in a plate heat exchanger

Auth and Loiano [18] report the following heat transfer correlation for plate heat exchangers: $Nu = a\, Re^{b} Pr^{c} (\mu/\mu_s)^{d}$, where a = 0.15 to 0.4, b = 0.65 to 0.85, c = 0.30 to 0.45 and d = 0.05 to 0.2. The characteristic dimension to be used in the Nusselt and Reynolds Numbers is the hydraulic mean (or equivalent) diameter D_e (see Section 2.1). μ is the fluid viscosity at the mean bulk temperature of the fluid and μ_s is the fluid viscosity at the mean wall temperature. It should be noted that the above equation applies to turbulent flow of fluid across the plate surface. This condition is normally met in plate heat exchangers since the design of the plate surface (which is corrugated rather than flat) promotes turbulence, with a critical Reynolds Number reduced to 150-300 (and in some cases as low as 10), compared with the usual value of 2,100. If the flow is laminar the following heat transfer correlation has been proposed [18]: $Nu = a\{Re\, Pr(D_e/L)\}^{0.33}(\mu/\mu_s)^{0.14}$, where L is the plate length and a has values of 1.86 to 4.5 depending on the plate characteristics. Manufacturers of plate heat exchangers will provide heat transfer correlations for each of their heat exchanger types.

Example 2.18

Calculate the overall heat transfer coefficient for heat exchange between two fluids flowing in countercurrent direction through a plate heat exchanger under the following conditions. Fluid 1 has a flow rate of 500 litres/hour and is to be heated from 10°C to 90°C. It has a viscosity of 1.70×10^{-3} kg/m s at 30°C and 0.60×10^{-3} kg/m s at 100°C, density of 900 kg/m^3, specific heat of 4000J/kgK and thermal conductivity of 0.5W/mK. Fluid 2 has a flow rate of 1500 litres/hour and enters at 95°C. It has a viscosity of 0.25×10^{-3} kg/m s at 100°C and 0.35×10^{-3} kg/m s at 80°C, density of 1000 kg/m^3, specific heat of 4200 J/kg K and thermal conductivity of 0.7W/mK.

The plates have dimensions 800mm x 200mm (effective surface area 0.15m^2) and are made of stainless steel of thickness 0.71mm, thermal conductivity 16.3W/mK, with a gap between plates of 2mm. Use the correlation $Nu = 0.20\ Re^{0.67} Pr^{0.33} (\mu/\mu_s)^{0.2}$.

(a) D_e = 4 x cross-sectional area/wetted perimeter = 4 x 0.2 x 0.002/{(2 x 0.2) + (2 x 0.002)} = 0.00400m. Velocity through channels = volumetric flow rate/cross-sectional flow area = {500/(1000 x 3600)}/(0.2 x 0.002) = 0.3472 m/s for fluid 1, and {1500/(1000 x 3600)}/(0.2 x 0.002) = 1.042 m/s for fluid 2.

(b) Mean bulk temperature of fluid 1 = (10 + 90)/2 = 50°C. If outlet temperature of fluid 2 is t_2, then enthalpy change in fluid 1 = enthalpy change in fluid 2 gives (500/1000 x 3600) x 900 x 4000 x (90 - 10) = (1500/1000 x 3600) x 1000 x 4200 x (95 - t_2). From this t_2 = 72.1°C. Therefore mean bulk temperature of fluid 2 = (95 + 72.1)/2 = 83.6°C.

(c) The mean wall temperature can be approximated as the average of the two mean bulk temperatures = (50 + 83.6)/2 = 66.8°C (the error in this approximation will usually have a negligible effect on the evaluation of $(\mu/\mu_s)^{0.2}$.)

(d) For fluid 1: μ(at 50°C) = 1.39×10^{-3} kg/m s and μ_s (69.7°C) = 1.076×10^{-3} kg/m s, assuming a linear relationship between viscosity and temperature over the range of temperatures involved.

Re_1 = 0.3472 x 0.00400 x 900/0.00139 = 899.

Pr_1 = 4000 x 0.00139/0.5 = 11.12; μ/μ_s = 1.39/1.076 = 1.292.

$Nu_1 = 0.20 \times 899^{0.67} \times 11.12^{0.33} \times 1.292^{0.2} = 44.41$

$h_1 = 44.41 \times 0.5/0.00400 = 5551\ W/m^2K$.

(e) For fluid 2: μ(at 83.6 °C) = 0.332×10^{-3} kg/m s. The viscosity correction term is negligible.

$Re_2 = 1.042 \times 0.00400 \times 1000/0.000332 = 12\ 554$

$Pr_2 = 4200 \times 0.000332/0.7 = 1.992$

$Nu_2 = 0.20 \times 12\ 554^{0.67} \times 1.992^{0.33} = 140$

$h_2 = 140 \times 0.7/0.00400 = 24\ 500\ W/m^2K$.

(f) The plate conductive resistance = $0.71 \times 10^{-3}/16.3 = 4.356 \times 10^{-5}$. Therefore the overall heat transfer coefficient (U) is given by $1/U = (1/5551) + (1/24\ 500) + 4.356 \times 10^{-5} = 2.645 \times 10^{-4}$. Hence U = 3780 W/m^2K.

Example 2.19

Calculate the number of plates required for the operation of the heat exchanger of example 2.18. What extra platage would be required to allow for deposits on the plate surfaces giving fouling resistances of $2 \times 10^{-4} m^2 sK/J$ on the fluid 1 side and $4 \times 10^{-5} m^2 sK/J$ on the fluid 2 side ?

(a) The plate surface area required = heat transferred (q)/U x log mean temperature difference across plates (Δt_{lm}). q = (500/1000 x3600) x 900 x 4000 x (90 - 10) = 40 000W. Δt_{lm} = (72.1 - 10) - (95 - 90)/ln (72.1/5) = 22.66 °C. Thus plate surface area = 40 000/ 3780 x 22.66 = 0.467 m^2. Therefore number of plates required = 0.467 /0.15 = 3.11. Hence 3 plates.

(b) The inclusion of fouling factors gives an increased value of $1/U = 2.645 \times 10^{-4}$ (original value) $+ 4 \times 10^{-5} + 2 \times 10^{-4} = 5.045 \times 10^{-4}$. Thus U = 1982 W/m^2K. The new platage area is therefore 40 000/ 1982 x 22.66 = 0.8906m^2 = 0.8906 / 0.15 plates = 5.94 plates. Hence a further 3 plates would be required.

One of the advantages of plate heat exchangers is that several heat exchange operations can be performed in a single unit. Figure 2.10 shows such an arrangement for a process liquid being initially heated (say to inactivate an enzyme) and then being cooled. Regeneration - preheating the incoming liquid by heat transfer from the outgoing system which needs to be cooled - can also be included.

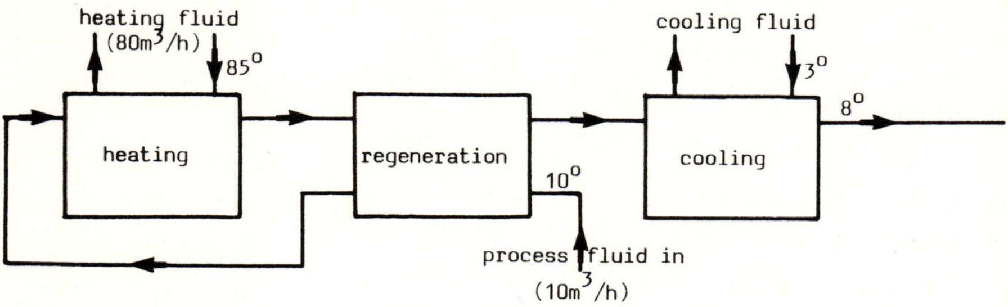

Figure 2.10 - Multiple heat exchange in a plate heat exchanger (see example 2.20)

Example 2.20

A plate heat exchanger is used for the duty shown diagramatically in figure 2.10. The process fluid is to be heated from 10°C to 80°C, in a maximum time of 20s, and then cooled to 8°C. The cooling liquid enters at 3°C and is to leave the exchanger at 5°C. 70% of the heating is to be carried out in the regeneration section. The process fluid has density 800 kg/m^3 and specific heat 4000 J/kgK. The heating and cooling fluids have density 1000 kg/m^3 and specific heat 4200 J/kgK.

The heat exchanger uses plates of dimension 1800 mm x 500 mm, effective surface area 0.7 m^2 with plate gaps of 1.5 mm. The overall heat transfer coefficients have been calculated to be 1200 W/m^{2}K in the regeneration section (U_r) and 2500 W/m^{2}K in the heating and cooling sections (U_h and U_c).

Establish a suitable plate arrangement for this duty.

(a) Regeneration section. The heat exchange required to heat process fluid from 10°C to 80°C = (10/3600) x 800 x 4000 x (80-10) = 6.222 x 10^{5}W. Therefore heat exchange in regeneration section (q_r) = 0.7 x 6.222 x 10^5 = 4.356 x 10^{5}W. Temperature change of process fluid in regeneration section = 0.7 x 70 = 49°C. Temperature of cold process fluid into section = 10°C; thus cold process fluid leaves at 59°C. Temperature of heated process fluid into regeneration section = 80°C; thus heated process fluid leaves at 31°C. Therefore temperature difference across plates in this section (Δt_r) = 80 - 59 = 31 - 10 = 21°C. Area of plates in this section = $q_r/U_r\Delta t_r$ = 4.356 x 10^5/1200 x 21 = 17.28m^2. Therefore number of plates = 17.28/0.7 = 25.

(b) Heating section. Heat exchange in this section (q_h) = 6.222 x 10^5 - 4.356 x 10^5 = 1.866 x 10^5W. Thus temperature change of heating liquid = q_h/mass flow rate x specific heat = 1.866 x 10^5/(80/3600) x 1000 x 4200 = 2.0°C. Temperature of heating liquid into section = 85°C; thus temperature out = 83°C. Temperature of process fluid in = 59°C and out = 80°C. The log mean temperature difference (Δt_h) = (83-59) - (85-80)/ln(24/5) = 12.11°C. Area of plates in this section = $q_h/U_h\Delta t_h$ = 1.866 x 10^5/2500 x 12.11 = 6.17m^2. Therefore number of plates = 6.17/0.7 = 9.

(c) Cooling section. Heat exchange in this section (q_c) = (10/3600) x 800 x 4000 x (31-8) = 2.044 x 10^5W. For the required 2°C rise in temperature of the cooling fluid, flow rate must be q_c/2 x 4200 = 24.34 kg/s = 87 524 l/hr. Therefore cooling water flow rate should be 90 000 litres/hour.

Temperature of process fluid into section = 31°C; out of section = 8°C. Thus log mean temperature difference = (31-5) - (8-3)/ln (26/5) = 12.74°C. Area of plates in this section = $q_c/U_c\Delta t_c$ = 2.044 x 10^5/2500 x 12.74 = 6.42m^2. Therefore number of plates = 6.42/0.7 = 9.

(d) Time for process fluid to reach 80°C. Number of plates used for heating = 25 (regeneration) + 9 (heating) = 34. Thus, number of plate gaps through which process liquid passes = 17. Distance travelled over one plate surface = 1.8 m. Therefore total distance travelled over plates during heating = 17 x 1.8 = 30.6 m. Gap cross-sectional area = 0.5 x 0.0015 = 0.00075m^2. Thus fluid velocity = (10/3600)/0.00075 = 3.704 m/s. Therefore time for process liquid to flow through 34 plates 30.6/3.704 = 8.3s. Even allowing for time to flow through ports etc., the heating time requirement of 20s maximum will be met.

In the above calculation the heat transfer coefficients have been stated - they could have been calculated as in example 2.18.

The other important consideration in plate heat exchanger specification is pressure drop. Pressure drop/flow rate correlations are specific to plate design and arrangement and data are supplied by the manufacturers for specific heat exchangers.

2.5 FREEZING

When food is frozen for preservation purposes it is important to obtain a rapid freezing rate, since it is known that slow freezing leads to irreversible damage to the food structure. Rapid freezing also allows increased outputs from the freezing equipment. A typical freezing process can be split into three parts (figure 2.11): a pre-cooling section where the product is brought down to its freezing point; a period during which ice-crystallisation occurs; and a tempering section where the product temperature falls from the freezing point towards the freezing medium temperature (note that figure 2.11 shows the temperature change with time at only one point in the product). The freezing point is lower than 0°C

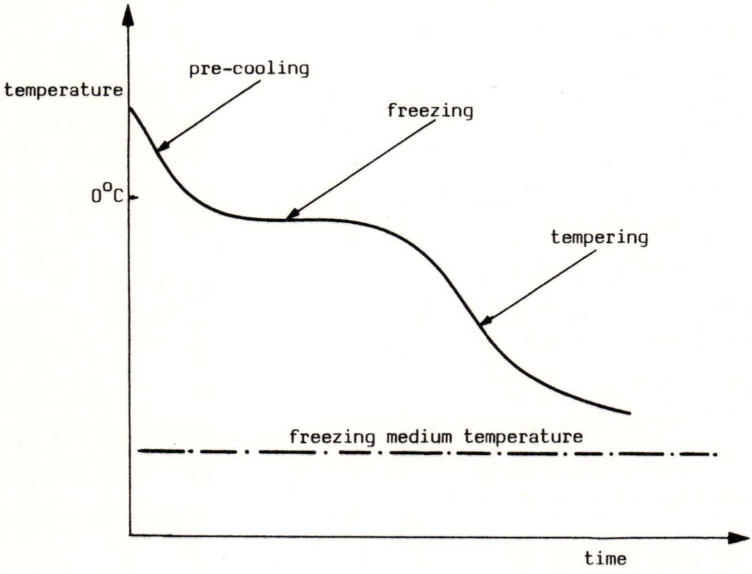

Figure 2.11 - Product temperature change with time during freezing

because of the presence of solutes in the liquid components of the food.

An exact determination of freezing time requires analysing the heat flow by unsteady state conduction through frozen and unfrozen sections of the food, as well as the heat exchange at the interface between these sections and at the surface of the product. Such analysis is complex and may be limited in application, particularly because of the lack of precise thermal property data for foods. Because of this other simpler models have been proposed but care needs to be taken in their use to make sure that the assumptions made are applicable to the particular freezing operation under investigation.

<u>Freezing time prediction by the Neumann method</u> [19]

The significant assumptions of the model are unsteady state conduction in the frozen and unfrozen sections and no heat transfer resistance at the product surface. Consider a very thick slab frozen from one major surface (which is at temperature t_m), with heat flowing only perpendicular to this surface. The time θ_f for a point in the slab, distance x from the surface, to fall from an initial temperature t_i to a temperature t_o is given by: $(t_o - t_m) = (t_f - t_m)\{\mathrm{erf}[x/2(a_1\theta_f)^{\frac{1}{2}}]/\mathrm{erf}\delta\}$, where δ is obtained from $\{\exp(-\delta^2)/\mathrm{erf}\delta\} - [\{k_2 a_1^{\frac{1}{2}}(t_i - t_f)\exp(-a_1\delta^2/a_2)\}/\{k_1 a_2^{\frac{1}{2}}(t_f - t_m)\,\mathrm{erfc}(\delta a_1^{\frac{1}{2}}/a_2^{\frac{1}{2}})\}] = \delta\, \lambda_f \pi^{\frac{1}{2}}/c_1(t_f - t_m)$. In these equations a_1 and a_2 are the thermal diffusivities of the frozen and unfrozen material, k_1 and c_1 are the thermal conductivity and specific heat of the frozen material, k_2 is the thermal conductivity of the unfrozen material, λ_f is the material latent heat, t_f is the materials freezing point, erf and erfc are the error and co-error functions.

This analysis is restricted in application since it describes the recession of the ice-front through a comparatively shallow layer of a deep body and therefore cannot accurately predict the freezing time of a finite body.

<u>Freezing time prediction by the Stefan method</u> [20]

If the material to be frozen is initially at its freezing point (but unfrozen), a considerable simplification can be made to the

Neumann solution given above. This is best expressed in dimensionless form [21] as $Fo_f = (Ko/2)(1+1/6\ Fo_f)$, where $Fo_f = a\theta_f/x^2$ and $Ko = \lambda_f/c_1(t_f-t_m)$. The assumption of zero heat transfer resistance (i.e. infinite heat transfer coefficient) is still made, but the solution can be applied to a finite thickness slab.

Freezing time prediction by the Plank method [21]

The significant assumptions of this model are steady state conduction in the frozen section and product initially at its freezing point (it can be shown [22] that for typical food properties the former condition will be met).

The equation derived by Plank [23] for the time for the slowest cooling point to be completely frozen (θ_f^1) is $\theta_f^1=\{\lambda_f\rho_1/(t_f-t_m)\}\{(Pd/h)+(Rd^2/k_1)\}$, where ρ_1 is the frozen product density and h is the heat transfer coefficient at the product surface. P and R are constants dependent on the product geometry: P = 1/6 for a sphere of diameter d, 1/4 for an infinite length cylinder of diameter d and 1/2 for an infinite length slab of thickness d frozen from its two major surfaces or thickness d/2 when frozen from one major surface. For these three shapes, R = 1/24, 1/16 and 1/8 respectively.

For a brick-shaped body the Plank equation takes the form [24]:

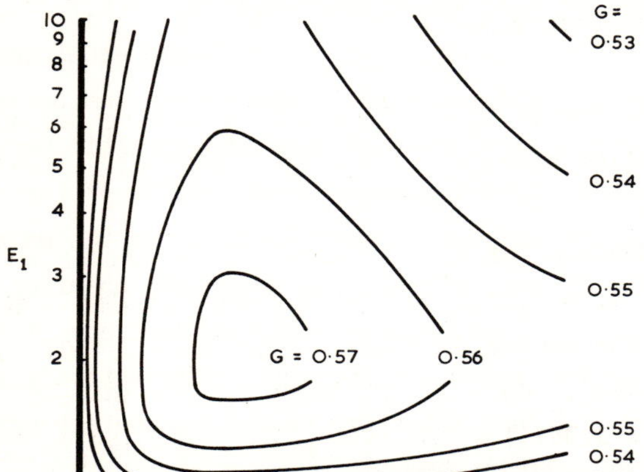

Figure 2.12 - Values of G for brick-shaped body of dimensions $d \times E_1 d \times E_2 d$ (from [21])

$\theta_f^1 = \{\lambda_f \rho_1 D/(t_f - t_m)\}\{(d/2h) + Gd^2/4k_1\}$, where D = 2 x volume of body/(area of cooled surface x shortest dimension of brick) and G can be obtained graphically (figure 2.12).

Freezing time prediction by modified Plank equation.

To allow for the initial temperature of the foodstuff being above the freezing point, Nagaoka [25] proposed a modification of the Plank equation, substituting a term λ_f^1 for λ_f, where $\lambda_f^1 = \{c_2(t_i - t_f) + \lambda_f\}\{1 + 0.008(t_i - t_f)\}$, using S.I. units, and c_2 is the specific heat of the unfrozen material. Cleland and Earle [26] have given a more complex form of modification.

The use of these methods for estimating freezing times is shown in the following examples.

Example 2.21

A comminuted meat product in sliced form (thickness 8 mm) is frozen by direct contact between refrigerated plates. The plates are at -30°C and the product properties are: k_1 = 1W/mK, c_1 = 1650 J/kgK, ρ_1 = 850 kg/m^3, k_2 = 0.35 W/mK, c_2 = 2800 J/kgK, ρ_2 = 950 kg/m^3, $\lambda_f = 3 \times 10^5$ J/kg, t_f = -3°C. Calculate the time to just freeze the centre of the product from an initial temperature of either 15°C or -3°C, with a surface heat transfer coefficient of 100W/m^{2}K or a negligible surface heat transfer resistance.

(a) Freezing time if t_1 = -3°C and zero surface resistance.

Plank's equation gives $\theta_f^1 = \{3 \times 10^5 \times 850/(-3 + 30)\}\{0 + 0.008^2/(8 \times 1)\}$ = 76s.

Stefan's equation for $t_1 = t_f$ requires Fo_f and Ko in the dimensionless form. $a_1 = k_1/c_1\rho_1$ = 1/1650 x 850 = 7.1301 x 10^{-7} m^2/s, and x = 0.004m for the centre position. Thus $Fo_f = 7.1301 \times 10^{-7} \times \theta_f^1/(.004)^2 = 0.044456\theta_f^1$. Ko = 3 x 10^5/1650 x (-3 + 30) = 6.734. Therefore $0.044456\theta_f^1 = 3.367(1 + 1/0.2674\theta_f^1)$; this gives θ_f^1 = 79s. Thus both methods are in close agreement.

(b) Freezing time if t_1 = 15°C and zero surface resistance.

For Nagaoka's equation, $\lambda_f^1 = \{2800 \times 18 + 3 \times 10^5\}\{1 + 0.008 \times 18\} = 4.009 \times 10^5$ J/kg. $\theta_f^1 = 4.009 \times 10^5 \times 850/27\}\{0 + 008^2/(8 \times 1)\}$ = 101s.

(c) Freezing time if $t_i = -3$ºC and h = 100 W/m^{2}K.
Plank gives $\theta_f^1 = \{3 \times 10^5 \times 850/27\}\{0.008/(2 \times 100) + 0.008^2/(8 \times 1)\}$ = 453s.

(d) Freezing time if $t_i = 15$ºC and h = 100 W/m^{2}K.
Nagaoka gives $\theta_f^1 = (4.009 \times 10^5 \times 850/27)\{0.008/2 \times 100 + 0.008^2/(8 + 1)\}$ = 605s.

Example 2.22

A foodstuff is packaged in box cartons with dimensions 5cm x 15cm x 15cm. The product properties are as in example 2.21. Calculate the time for the centre of the product to be frozen in a high velocity air freezer if the product is initially at 15ºC and the surface heat transfer coefficient is 200 W/m^{2}K.

The equation for freezing time for a brick-shaped body includes the shape factors D and G. The carton volume is 0.05 x 0.15 x 0.15 = 0.001125m^3; cooled surface area = $2 \times 0.15^2 + 0.05 \times 0.60 = 0.075$m^2. Thus D = 2 x 0.001125/0.075 x 0.05 = 0.6. From figure 2.12, for $E_1 = E_2 = 3$, G = 0.567. Using λ_f^1 to allow for pre-cooling, $\lambda_f^1 = \{2800 \times 18 + 3 \times 10^5\}\{1 + 0.008 \times 18\} = 4.009 \times 10^5$ J/kg and $\theta_f^1 = \{4.009 \times 10^5 \times 850 \times 0.6/27\}\{0.05/400 + 0.567 \times 0.05^2/4 \times 1\}$ = 3630s = 60.5min.

Example 2.23

A fruit purée at 6ºC is poured into deep metal trays which are placed on refrigerated plates, the surfaces of which are kept at a constant -30ºC. The thermal conductivity, specific heat and density of the frozen puree are 1.5 W/mK, 1800 J/kgK and 880 kg/m^3 respectively and 0.4 W/mK, 3200 J/kgK and 970 kg/m^3 for the unfrozen purée. The latent heat of freezing is 3×10^5 J/kg and the freezing point of the purée is -4ºC. Use Neumann's equation to predict the thickness of product which would be frozen after 10 minutes.

Thermal diffusivity of frozen product = 1.5/(1800 x 880) = 9.470 x 10^{-7} m^2/s and similarly for the unfrozen product = 1.289 x 10^{-7} m^2/s.

First, calculate δ. This is obtained from $\{\exp(-\delta^2)/\text{erf}\ \delta\} - [\{0.4 \times (9.470 \times 10^{-7})^{\frac{1}{2}} \times (6 + 4) \times \exp(-9.470 \times 10^{-7}\delta/1.289 \times 10^{-7})\}/\{1.5 \times (1.289 \times 10^{-7})^{\frac{1}{2}} \times (-4 + 30) \times \text{erfc}\ (9.470/1.289)^{\frac{1}{2}}\delta\}] = \delta \times 3 \times 10^5 \times \pi^{\frac{1}{2}}/1800 \times 26$. Substituting values of δ into this equation until L.H.S. = R.H.S. gives the value of δ = 0.252. Values of the error function are given in table 2.1 (erfc X = 1 - erf X).

Table 2.1 - Error function values

X	erf X	X	erf X
0	0	0.5	0.52050
0.1	0.11246	0.8	0.74210
0.2	0.22270	1.0	0.84270
0.25	0.27633	1.5	0.96611
0.3	0.32863	2.0	0.99532
0.4	0.42839	2.5	0.99959

Thus $(-4 + 30) = (-4 + 30)\ \{\text{erf}[x/2(9.470 \times 10^{-7} \times 600)^{\frac{1}{2}}]/\text{erf}\ 0.252\}$, where x is the thickness frozen in 10 minutes. This gives the thickness frozen as 1.20 x 10^{-2}m, i.e. 1.2 cm.

Example 2.24

Joints of meat, of approximately spherical shape and with a maximum thickness of 18 cm, are found to have a freezing time of 5 hours at a freezing medium temperature of -30°C. (a) How long would it take to freeze 15 cm diameter joints in the same freezer ? (b) Would it be worthwhile increasing the air velocity in the freezer to obtain faster freezing ? (c) How much would freezing at -40°C reduce the freezing time ?

(a) If heat transfer to the freezing medium is mostly controlled by conduction within the product, Nagaoka's equation shows freezing time $\propto d^2$(since Pd/h is negligible); if controlled by convection freezing time $\propto$ d. In this example the freezing material is large and internal conduction will probably control the freezing time. Therefore freezing time for 15 cm joint will be approximately 5 hours x $(15/18)^2$ = 3½ hours. More precisely, a likely value for h is about 150 W/m^3K and k_1 for meat is approximately 1 W/mK.

Thus $(Pd/h + Rd^2/k_1)$ for the larger joint = $0.18/(6 \times 100) + 0.18^2/(24 \times 1) = 0.00165$; for the smaller joint this term is $0.15/(24 \times 1) = 0.0011875$. Since freezing time is proportional to this term the smaller joint has a freezing time = $(0.0011875/0.00165) \times 5$ hrs = 3.6 hours.

(b) Since Pd/h is much smaller than Rd^2/k_1 the effect of a change in air velocity, and hence h, on freezing time will be small. Doubling the velocity would increase h by a factor of about 1.74 (taking $h \propto \text{velocity}^{0.8}$, see Chapter 2). An increase of h from 100 to 174 would decrease $(Pd/h + Rd^2/k_1)$ from 0.00165 to 0.0015224 and this only reduces freezing time to 92% of its original value.

(c) The freezing time is inversely proportional to $(\theta_f-\theta_m)$ and thus reducing the freezing medium temperature to -40°C would reduce the freezing time to 5 hrs x $(-3 + 30)/(-3 + 40) = 3.65$ hours, assuming a freezing point of -3°C.

Duration of the tempering period.

Rutov [27] estimates the time for a product to cool to a centre temperature t_o from a freezing point t_f as $\{2nd^2/\pi^2a_1\}\{\ln[(t_f-t_m)/(t_o-t_m)] -0.21\}\{(2k_1/hd)+1/2\}$, where n = 1.03 - 1.06 for quick freezing and 1.16 for slow freezing.

Example 2.25

The comminuted meat product of example 2.21 is to be cooled to a centre temperature of -25°C. Calculate this cooling time for a heat transfer coefficient of 100 W/m^2K.

Taking n = 1.05, cooling time = $\{2 \times 1.05 \times 0.008^2/\pi^2 \times 7.1301 \times 10^{-7}\}\{\ln(27/5) - 0.21\}\{(2 \times 1/100 \times 0.008) + 0.5\} = 85s$.

Fluidised bed freezing

Food pieces can be frozen in fluidised beds by blowing refrigerated air through the pieces. If the heat transfer coefficient is known (see Chapter 7) the freezing time can then be calculated by one of the above equations, and hence the throughput of the fluidised bed freezer.

Example 2.26

The peas of example 7.4 are frozen in a fluidised bed of 4m length, 1m width with air at -30°C. The bed depth is 5 cm. Calculate the throughput of the freezer if the peas enter at 15°C and have a freezing point of -2°C. ($\lambda_f = 3 \times 10^5$ J/kg, c_2 = 3600 J/kgK, k_1 = 0.4 W/mK.)

From example 7.4, h = 198 W/m^{2}K. For Nagaoka's equation, λ_f^1 = $[3600 \times 17 + 3 \times 10^5][1 + 0.008 \times 17] = 4.103 \times 10^5$ J/kg. Therefore $\theta_f^1 = (4.103 \times 10^5 \times 950/28)\{(0.008/6 \times 198) + (0.008^2/24 \times 0.4)\}$ = 187s. For this residence time in the freezer, bulk volume throughput = bed depth x bed area/residence time = 0.05 x 4/187 = $1.069 \times 10^{-3} m^3/s$. 1m^3 of bed has solids volume = $(1-\varepsilon)m^3 = 0.55m^3$ and mass of 0.55 x 950 kg. Therefore freezer throughput = $1.069 \times 10^{-3} \times 0.55 \times 950$ = 0.559 kg/s = 2010 kg/hr.

2.6 COLD STORE DESIGN

Cold storage refers to the holding of chilled or frozen product in insulated rooms which are kept at a controlled temperature by extraction of heat through a refrigeration system. This section deals with the calculation of the thermal load to be extracted in such a store. Details of the refrigeration systems are given in references [28, 29].

The sources of thermal load are:

(a) enthalpy change of product. Ideally product going into a cold store should be at the same temperature as the cold store, and there will be no enthalpy change. If the product is cooled the required average rate of heat extraction will be $mc(t_i-t_o)/\theta_s$, where m is the mass of product, c is product specific heat, t_i is the temperature of product entering the store, t_o is the product temperature after storage for time θ_s (t_o can be calculated from unsteady state heat transfer analysis as described in Chapter 2.2). If product freezes during storage the heat to be extracted will be $m\{c_2(t_i-t_f) + \lambda_f + c_1(t_f-t_o)\}$, where t_f is the product freezing point, λ_f is the latent heat of freezing and subscripts 1 and 2 refer to the frozen and unfrozen material.

(b) heat of respiration of product. Fruits and vegetables evolve carbon dioxide during storage and this respiration produces an associated evolution of heat. Desrosier [30] gives values of heat of respiration (E_1) for number of foodstuffs, some of which are given in Table 2.2. The required rate of heat extraction is mE_1.

Table 2.2 - Heats of respiration (J/kg s) at:

	0^oC	15^oC
Apples	0.004 - 0.010	0.027 - 0.042
Green beans	0.066 - 0.074	0.386 - 0.530
Carrots	0.026	0.097
Peas	0.098	0.472
Pears	0.008 - 0.011	0.106 - 0.159
Potatoes	0.005 - 0.011	0.026 - 0.042

(c) Cooling of racks, containers etc. This will require a rate of heat extraction of $m_r c_r (t_r - t_c)/\theta_s$, assuming the racks etc, which are usually metal with a high thermal conductivity, reach the cold store temperature t_c (subscript r refers to the properties of the racks etc.).

(d) Heat output from electrical fittings (lighting etc.) Since a large percentage of this electrical energy is converted to thermal energy the required rate of extraction for this source can be approximated by the wattage of the devices.

(e) Heat evolved by workpeople in the store. The heat output from the human body has been established for a number of activities; a suitable figure for work involved in cold storage operation is 8.5×10^5 J/man hour.

(f) Heat flow through cold store walls, roof and floor. Heat flows through the insulation of the walls and roof of the store at a rate of $k_w A_w (t_a - t_c)/x_w$, where k_w is the wall material thermal conductivity, A_w is the wall and roof surface area, t_a is the external temperature, x_w is the wall thickness. Heat flow through the floor is also by conduction, but the thermal conductivity will be that of the structural materials of the floor (usually concrete). The external temperature below the floor depends on the method of cold store construction. If the store is built directly onto the ground and is operating below freezing point the ground temperature is commonly kept at a temperature just above freezing point of water by the insertion of heating elements - this eliminates the problem of damage to the floor by 'frost-heave' caused by the freezing of water in the floor structure.

(g) Heat ingress through door openings. The opening of doors into the cold store can allow substantial ingress of air at the external temperature into the store, its associated heat content having to be extracted by the refrigeration system. A number of systems are used for reducing the air exchange (e.g. air curtains, where air is blown across the door opening; double doors forming an air-lock between sets of doors into the store; self-closing flexible flap doors). Table 2.3 gives figures for the number of air changes per day occurring in typical cold store operation, from which the heat ingress can be calculated as 8.5×10^4 J/m^3 of store volume for each air change.

Table 2.3 - Air changes per 24 hours in typical cold store operation (from [31])

Volume of store (m^3)	No. of air changes
8	30
30	13
50	11
100	7.5
500	3
1000	2.1
3000	1.1

The total of all the above thermal loads gives the total rate of heat evolution in the cold store. The refrigeration system has to extract heat at at least this rate. If the rate of evolution and rate of extraction were identical the refrigeration unit would run continuously. This would be poor engineering practice and therefore it should be designed to extract at a faster rate, the refrigeration unit being switched on and off via a thermostat in the cold store. A reasonable design figure for the percentage 'on' time of the refrigeration unit is 75%.

The rate of heat extraction by refrigeration systems is commonly expressed in 'tons of refrigeration', where 1 ton = 3517W.

It should be noted that there can be considerable fluctuations in the thermal load in a typical store, particularly if it is subjected to very active periods of loading and unloading. It is therefore sensible to check the thermal load under the most severe operating conditions as well as for average operation.

Example 2.27

A cold store, of dimensions 4m height, 8m width, 15m length, and operating at 2°C, is to be used for storing harvested fruit prior to processing. The average period in the cold store is 24 hours, during which time the fruit cools from 16°C to 8°C. The fruit enters, and is removed from, the cold store at an approximately regular rate of 400kg/hour and at any time the average amount of product in the store is 4000kg.

Calculate the refrigeration duty for this cold store given the following data. The fruit has a specific heat of 3600J/kgK and an average respiration rate of 0.05J/kg s between 16°C and 8°C. The ambient temperature outside the cold store is 13°C. The cold store walls and roof are made of 130mm expanded polystyrene insulation (thermal conductivity 0.035 W/mK). The cold store floor is 500mm thick, with a thermal conductivity of 0.06 W/mK, on ground at 8°C. Product enters the store on metal racks of mass 1000kg/1000kg of product (specific heat of the metal is 500 J/kgK). The other thermal loads are two workmen in the store for 2 hours/day and four 200W lights for 4 hours/day. Assume four air changes per day.

Thermal loads:

(a) product cooling: $mc(t_i - t_o)/\theta_s$ = (400 x 24) x 3600 x (16-8)/ 24 x 3600 = 3200W

(b) heat of respiration: 4000 x 0.05 = 200W

(c) cooling racks: (400 x 24) x 500 x (16-2)/24 x 3600 = 778W (assuming racks enter at product temperature and cool to cold store temperature)

(d) electrical fittings: 4 x 200 x (4/24) = 133W

(e) workmen: at 8.5×10^5 J/man hour = 8.5×10^5 x (2 x 2)/

24 x 3600 = 39W

(f) through walls and roof: area of walls and roof = (4 x 46) + (8 x 15) = $304m^2$, therefore, heat flow = 0.035 x 304 x (13-2)/0.13 = 900W

through floor: 0.06 x 120 x (8-2)/0.5 = 86W

(g) air changes: volume of store = 4 x 8 x 15 = $480m^3$. There are 4 air changes per day = 4.6296 x 10^{-5}/sec. Heat flow = 8.5 x 10^4 J/m^3 of store volume for each air change = 8.5 x 10^4 x 480 x 4.629 x 10^{-5} = 1889W.

Thus, total load = 3200 + 200 + 778 + 133 + 39 + 900 + 86 + 1889 = 7225W.

Allowing for refrigeration on for 75% of the time, required refrigeration duty is 7225 x 100/75 = 9.63kW = 2.74 tons of refrigeration.

Example 2.28

It is proposed to alter the loading procedure of the store in example 2.27, with 3000kg of product being taken into the empty cold store over a period of one hour. Determine whether this would be a suitable procedure.

It is necessary to establish whether the thermal load will be greater than the refrigeration capacity of the unit. The initial period of loading and product cooling will give the greatest thermal load - therefore calculate the load in the first two hour period (including the loading time).

(a) product cooling: an estimate of product temperature after two hours is required, which can be obtained by unsteady state heat transfer analysis (Chapter 2.2). An average temperature of 13°C is taken here. Then heat load = 3000 x 3600 x (16-13)/2 x 3600 = 4500W.

(b) heat of respiration: since the temperature is high in this initial period the heat of respiration is high (see table 2.2). Taking a value of 0.15J/kgs gives heat load = 3000 x 0.15 = 450W.

(c) cooling racks: 3000 x 500 x (16-2)/2 x 3600 = 2917W, since it is assumed that the thermal conductivity of the metal is sufficiently high for the racks to cool to cold store temperature over the two hour period.

(d) electrical fittings: assume lighting will be on throughout the loading period, then load = 4 x 200 x (1/2) = 400W.

(e) workmen: assume workmen will be in the cold store for one hour of this period, then load = 8.5 x 10^5 x 2 x (1/2)/3600 = 236W.

(f) through walls, roof and floor: 986W (as before)

(g) air changes: the loading period will lead to much greater air ingress - take a rate of 10 air changes/day for the whole two hour period. Then heat load = 8.5 x 10^4 x 48 x (10/240 x 3600) = 4722W.

The total of the above is 14.2kW. This is well in excess of the 9.63kW refrigeration duty calculated in example 2.27. If a refrigeration unit of 9.63kW capacity was used the proposed loading procedure would lead to a rise in temperature in the cold store even with the refrigeration continuously on. An estimate of the rate of rise of temperature can be made as follows. If 14.2kW are produced in the store and 9.6kW are being extracted, the difference (4.6kW) will cause the temperature in the store to rise. If the volume of air in the store is half the store volume (i.e. 0.5 x 480m^3) then the mass of air in the store = 240 x 1.2 = 288kg, taking air density = 1.2kg/m^3. Rate of temperature rise of the air = heat input/mass x specific heat = 4600/288 x 1000 = 0.0160°C/s = 0.96°C/min, taking air specific heat = 1000J/kgK. A potential temperature rise of 1°C/minute is excessive and for a refrigeration unit of this size this loading procedure would not be recommended.

REFERENCES

1. T.B. Drew, H.C. Hottel and W.H. McAdams, 'Heat transmission', Trans.Am.Inst.Chem.Engrs, 32 (1936) 271
2. E.N. Sieder and G.E. Tate, 'Heat transfer and pressure drop of liquids in tubes', Ind.Engng Chem., 28 (1936) 1429
3. F.W. Dittus and L.M.K. Boelter, 'Heat transfer in automobile radiators of the tubular type', Univ.Calif.Publs Engng, 2 (1930) 443
4. A.P. Colburn, 'A method of correlating forced convection heat

transfer data and a comparison with fluid friction', Trans.Am. Inst. Chem.Engrs, 29 (1933) 174

5. T.H. Chilton, T.B. Drew and R.H. Jebens, 'Heat transfer coefficients in agitated vessels', Ind.Engng.Chem., 36 (1944) 510
6. G.H. Cummings and A.S. West, 'Heat transfer data for kettles with jackets and coils', Ind.Engng Chem., 42 (1950) 2303
7. R.A. Bowman, A.C. Mueller and W.M. Magle, 'Mean temperature difference in design', Trans.Am.Soc.Mech.Engrs, 62 (1940) 283
8. H.S. Carslaw and J.C. Jaeger, Conduction of heat in solids, 2nd Ed., (Clarendon Press, Oxford, 1959)
9. P.J. Schneider, Temperature response charts, (Wiley, New York, 1963)
10. R.L. Pigford, 'Nonisothermal flow and heat transfer inside vertical tubes', Chem.Engng Prog.Symp.Ser., 51 (1955) 79
11. A.B.Metzner, Advances in Heat Transfer, Vol.2, p.357, (Academic Press, London and New York, 1965)
12. A.H.P. Skelland, Non-Newtonian Flow and Heat Transfer, (Wiley, New York, 1967)
13. A.B. Metzner and D.F. Gluck, 'Heat transfer to non-Newtonian fluids under laminar-flow conditions', Chem. Engng Sci.,12 (1960) 185
14. E.B. Christiansen and S.E. Craig, 'Heat transfer to pseudoplastic fluids in laminar flow', A.I.Ch.E.Jl, 8 (1962) 154
15. E. Hirai, 'Theoretical explanation of heat transfer in laminar region of Bingham fluid', A.I.Ch.E.Jl, 5 (1959) 130
16. L. Schiller, 'Die Entwicklung der laminaren Geschwindigkeitsverteilung und ihre Bedeutung Fur Zahigkeitsmessungen', Z. angew. Math. Mech., 2 (1922) 96
17. D.C. Bogue, 'Entrance effects and prediction of turbulence in non-Newtonian flow', Ind.Engng Chem., 51 (1959) 874
18. W.J. Auth and J. Loiano, in Practical aspects of heat transfer, (A.I.Ch.E., New York, 1978)
19. F. Neumann, cited by H.S. Carslaw and J.C. Jaeger, Conduction of heat in solids, 2nd Ed., (Clarendon Press, Oxford, 1959)
20. J. Stefan, Annln.Phys.u.Chem.,42 (1891) 269
21. N.D. Cowell, 'The calculation of food freezing times', Proc.12th Intern. Congr. Refrig., 2 (1967) 667

22. J. Lamb and E.Kinder, 'The prediction of freezing times of foodstuffs' in Meat Freezing - Why and How?, M.R.I. Symp.3, (Meat Research Institute, Bristol, 1974)
23. R. Plank, Z.ges.Kalteind., R3 (1941) H10
24. A.J. Ede, 'The calculation of the freezing and thawing of foodstuffs', Mod. Refrig. Air Control, 52 (1949) 52
25. J. Nagaoka et al, 'Experiments on the freezing of fish in an air-blast freezer', Proc.9th Intern.Congr.Refrig., 2 (1955) 4
26. A.C. Cleland and R.L. Earle, 'A comparison of analytical and numerical methods of predicting freezing times of foods', J.Fd.Sci., 42 (1977) 1390
27. D.G. Rutov, Proc.7th Intern.Congr.Refrig., 4 (1936) 211
28 W.R. Woolrich, Handbook of Refrigerating Engineering, Vols.1 and 2, 4th Ed., (AVI, Westport, 1965/6)
29. S.M. Henderson and R.L. Perry, Agricultural Process Engineering, 3rd Ed.,(AVI, Westport, 1976)
30. N.W. Desrosier, The Technology of Food Preservation, 3rd Ed., (AVI, Westport, 1970)
31. W.R. Woolrich and E.R. Hallowell, Cold and Freezer Storage Manual, (AVI, Westport, 1970)
32. A.S. Foust et al, Principles of Unit Operations, (Wiley, New York, 1960)
33. ibid, 2nd Ed., (Wiley, New York, 1980)

3 STERILISATION PROCESSES

3.1 CANNING AND BOTTLING

A wide range of foods can be preserved by the process of thermal sterilisation. In its most common form this involves placing the food in a suitable container, closing this in such a way that no micro-organisms can enter it, and then raising its temperature to a sufficiently high level and for a sufficient time for micro-organisms within the food to be effectively destroyed. This is called in-can sterilisation, since the most common container is the tin-plated steel can. Some micro-organisms are very resistant to destruction even at high temperatures and complete sterilisation of most foodstuffs would cause unacceptable thermal degradation of the product (e.g. seriously damaged texture or excessive loss of nutrients). It is only necessary to destroy those micro-organisms which are dangerous to man (pathogenic) and those which would lead to degradation of the food by microbiological attack. Indeed, in an exact scientific sense, it is not possible to obtain total destruction of even these micro-organisms since it is known that micro-organisms show a logarithmic decrease in the number of survivors as they are subjected to increasing heating times at any particular temperature. This means that there is an asymptotic approach to zero survivors with increasing processing time. It is therefore necessary to define 'commercial sterility' on the basis of commercial experience of percentages of contaminated processed cans and a knowledge of the distribution and growth characteristics of micro-organisms. A commonly used standard for commercial sterility is a probability of survival of Clostridium botulinum spores (a particularly dangerous micro-organism) of 1 in 10^{12}. Further information on the science and technology of canning will be found in references [1] and [2].

It is necessary to have a procedure for determining whether the canning process for a particular product will ensure adequate sterilisation. There are two commonly used techniques for this - the General Method and the Mathematical Method.

<u>General Method.</u>

The steps in this can be summarised as follows:

(a) Assuming that there is a logarithmic relationship between number of surviving micro-organisms and holding time (at any one temperature), the time for a ten-fold reduction in the number of survivors for the micro-organism of concern is determined at a number of temperatures. This is known as the decimal reduction time (D).

(b) A plot of log D against temperature is generally found to be a straight line. From such a plot the temperature change for a ten-fold change in D is established. This is known as the z value. For <u>Clostridium botulinum</u> it is 10°C(18°F).

The above two steps are in the province of the microbiologist - they define the characteristics of the micro-organism and the information they give can be used to analyse the effect of any sterilisation process on the survival of the particular micro-organism defined growing in a particular foodstuff.

(c) Given the z value, the relative rate of destruction of a micro-organism at different temperatures can be determined. The most common base temperature for comparison of destruction rates is 121.1°C(250°F). If the rate at 121.1°C is arbitrarily defined as 1, then from the definition of z, the rate at another temperature t is given by $L_v = 10^{-(121.1-t)/z}$ e.g. for a z value of 10°C, L_v (the lethal rate) is 0.1 at 111.1°C and 0.01 at 101.1°C.

(d) For the process under investigation a graph is plotted of temperature at the slowest heating point in the can against process time - a heat penetration curve. Typically it will have a shape as shown in figure 3.1. For any particular time on this graph the lethal rate can be computed from the temperature. This allows a graph to be drawn of lethal rate against process time, as in figure 3.2.

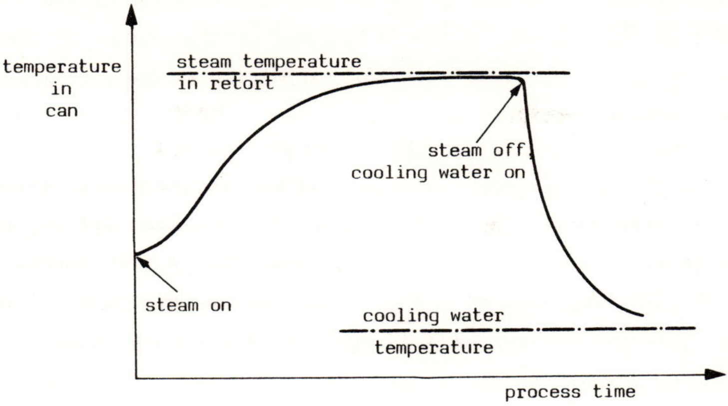

Figure 3.1 - Typical heat penetration curve in canning

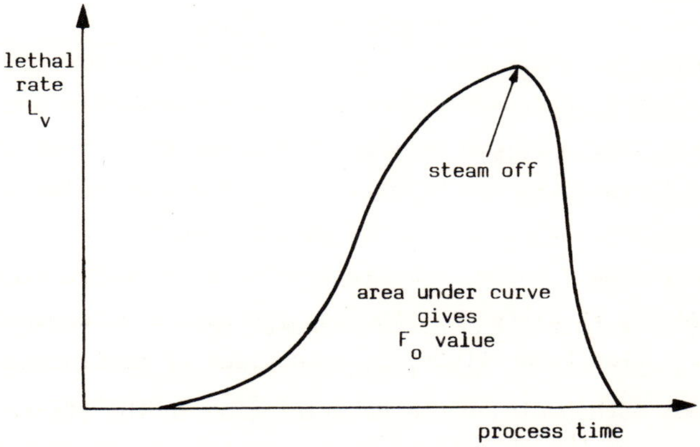

Figure 3.2 - Typical graph of lethal rate against process time

(e) The <u>total</u> destruction of the micro-organisms over the whole process is

$$\int_{\text{start of process}}^{\text{end of process}} L_v . d\theta, \text{ where } \theta \text{ is time in minutes.}$$

This integration cannot be performed analytically since the exact mathematical relationship between L_v and θ is not known. A graphical integration can be used by measuring the area under the curve. The value of the integral is called the F_o value of the process.

It should be noted that it is not an absolute measure of the lethality of the process, since it depends on the choice of the base-line temperature (121.1°C), the definition of lethal rate as unity at that temperature, and a value of z = 10°C. It has dimensions of time and is nearly always quoted in minutes.

(f) The F_o value allows comparison of the relative sterilisation effects of different processes. Two processes giving different heat penetration curves will have the same sterilisation effect if they produce the same F_o value. More importantly the F_o value obtained in a process can be compared with the values found to be necessary to ensure commercial sterility in particular packs of canned food. Some typical required F_o values are given in table 3.1. [3].

Table 3.1 - F_o values for commercial sterility of some food packs (minutes)

Carrots in brine	:	3-4
Cream soups	:	4-5
Evaporated milk	:	5
Peas in brine	:	6
Meats in gravy	:	12-15
Pet foods	:	15-18

The higher required F_o values in some packs are needed because they can contain highly heat resistant micro-organisms which, although not pathogenic, cause deterioration in the stored product. The lower values are typical of those for the destruction of Clostridium botulinum. For highly acid products (say below pH 3.9), such as fruit packs, it is not necessary to use processing temperatures of higher than 100°C since there is no multiplication of significant micro-organisms under these conditions. These products, which are often packed in bottles rather than cans, have effectively zero F_o values.

Example 3.1

Table 3.2 shows the centre temperature in a can of a food product sterilised in a steam retort operating at 120°C. Calculate (a) the F_o value of the process and (b) the process time required to achieve an F_o value of 4 in this pack. Assume a z value of 10°C.

Table 3.2

Time (mins)	Temperature (°C)	Time (mins)	Temperature (°C)
0 (steam on)	75.1	30 (steam off)	119.6
2.5	81.2	cooling water on)	
5	92.2	32.5	117.6
7.5	101.1	35	107.8
10	108.6	40	82.1
15	114.7	50	51.5
20	117.6	60	36.0
25	119.0	70	30.0

(a) Plot the heat penetration curve as in figure 3.3. This

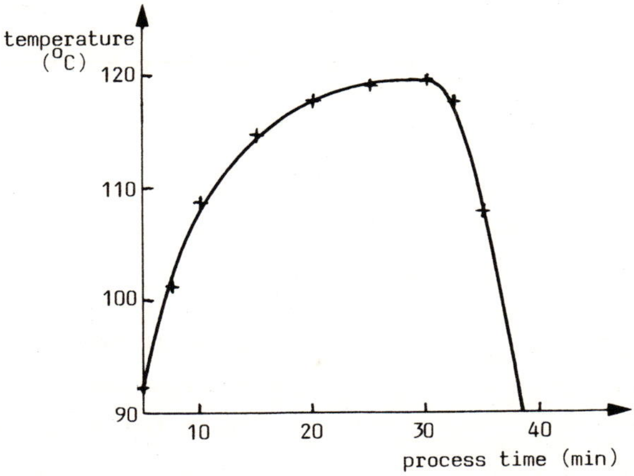

Figure 3.3 - Heat penetration curve for example 3.1

needs to be drawn accurately to obtain a satisfactory lethal rate curve and therefore only the time period giving significant lethal rate values has been covered, to enlarge the scale. Using the data from the curve (not the raw data) compute L_v values from $L_v = 10^{-(121.1-t)/z}$ and plot the lethal rate graph - as in figure 3.4.

Calculating the area under the curve (e.g. by counting squares) and dividing by the area of one F_o unit (e.g. 10 minutes x 0.1 L_v) gives the F_o value of the process. This is 10.6 (minutes).

(b) To achieve an F_o value of 4 the process time has to be

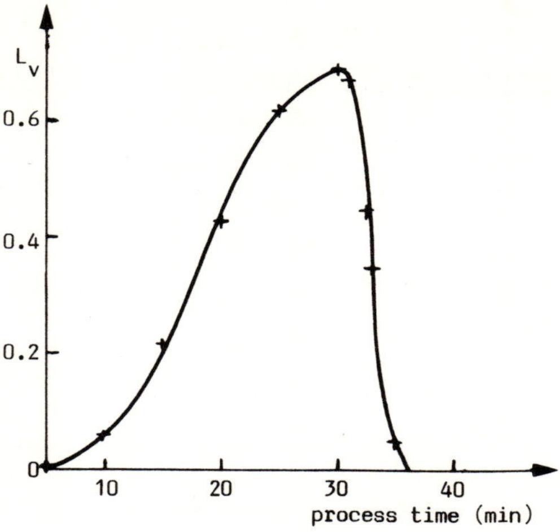

Figure 3.4 - Lethal rate graph for example 3.1

reduced. This means turning off the steam to the retort at an earlier stage - up to that point the heat penetration curve will be the same as in (a). The new cooling curve for the can will be of equal gradient to that in (a) providing that there is approximately the same temperature difference between the can contents and the cooling water at the start of the cooling period. If this is the case it is then necessary to locate the cooling curve which will lead to a bounded area of 4 units. The easiest way to find this is to draw a number of cooling lines (parallel to the original) starting at arbitrarily chosen times (this is shown in figure 3.5) and measure the bounded area in each case.

This leads to the data given in table 3.2.

Table 3.2

Conditions at start of cooling			Area under curve
Time (mins)	Temperature (°C)	L_v	(F_o)
30	119.8	0.741	10.6
26.5	119.5	0.69	8.2
23.5	118.8	0.585	5.3
17	115.9	0.30	1.76

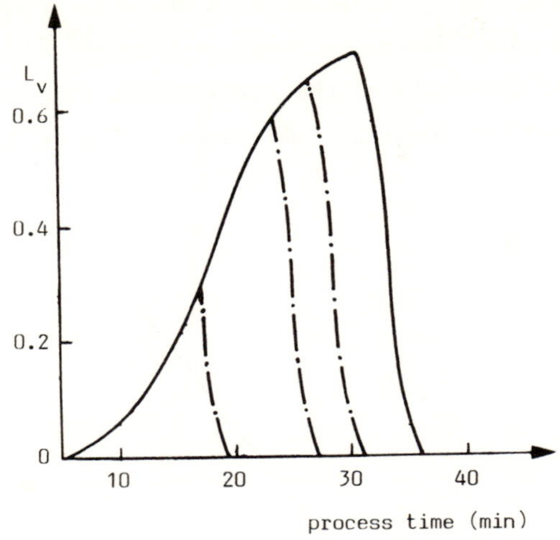

Figure 3.5 - Lethal rate graph with different cooling curves for example 3.1

From a graph of F_0 against time from table 3.2 the time for an F_0 value of 4 can be determined; this is 21.7 minutes. Note that the temperature difference between the centre can temperature at the end of heating and the cooling water (say at 20°C) is approximately the same for both the F_0 = 4 and F_0 = 10.6 cases and therefore the assumption of parallel cooling curves is valid.

Mathematical Method

This technique is based on an assumption that there will be a logarithmic approach of the can centre temperature to the retort temperature during the heating process (see Section 2.2 for the basis of this assumption). A full treatment of the method is given by Stumbo [2]. The steps can be summarised as follows:

(a) Plot the heat penetration curve for the product during heating as log (temperature deficit) i.e. log (retort temperature [R.T.] - can centre temperature) against time, with the log scale increasing downwards. As shown in figure 3.6 this should give a straight-line relationship for most of the heating time, with an initial curved log period. The 'zero' of the time scale makes allowance for the time delay between steam entering the retort and the interior of the retort coming up to steam temperature (the 'come-up-time' or C.U.T.). The usual procedure based on theoretical and experimental evidence is to allow 42% of the come-up-time as part of the heating period. This gives the zero time shown in figure 3.6.

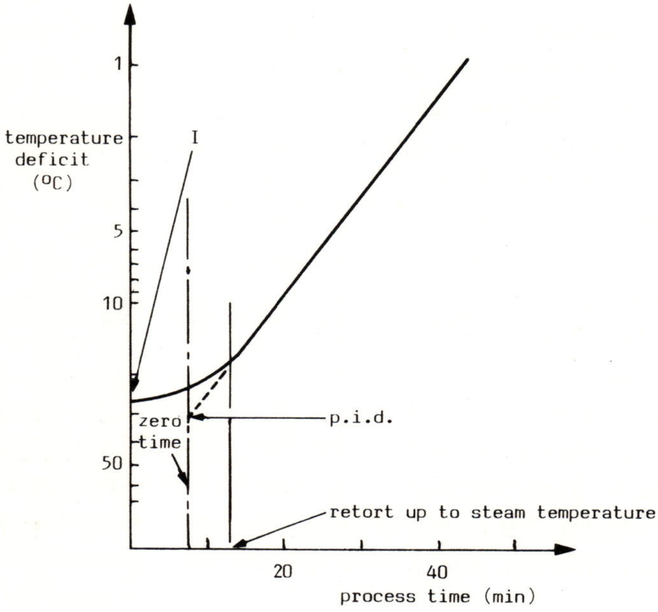

Figure 3.6 - Heat penetration curve for Mathematical Method analysis

(b) Extrapolate the straight line portion of the curve and determine the following four values: (i) heat penetration factor (f_H) = time for the straight-line plot to traverse one log cycle of temperature deficit; (ii) initial temperature deficit (I); (iii) pseudo initial deficit (p.i.d.) = temperature deficit at zero time given by intercept of the extrapolated straight line portion of the curve; (iv) j value = ratio of p.i.d. to I.

Having found f_H, j and I the values can be substituted in the equation of the straight-line heat penetration curve $B = f_H(\log jI - \log g)$, where g is the temperature deficit after time B.

Thus this equation gives the process time to achieve a particular temperature deficit. Therefore in order to determine the process time for the desired sterilisation it is necessary to know the required temperature deficit at the end of heating.

(c) Calculate F, the number of minutes at 121.1°C required for destruction of micro-organisms to the level desired in the process. By definition $F = D_{121.1}(\log a - \log b)$, where a and b are the number of micro-organisms at the beginning and end of the process.

(d) Calculate the required temperature deficit at the end of heating from tabulated data relating g to f_H/U, where U is the number of minutes for sterilisation at the retort temperature (R.T.) i.e. $U = F.10^{(121.1-R.T.)/z}$. It has been found that the relationship between f_H/U and g depends not only on the sterilisation characteristics of the micro-organisms as given by D and z but is also affected by the characteristics of the cooling curve. This is expected since it has been seen that a substantial proportion of the sterilisation of the product takes place during cooling. The significance of the cooling effect is best characterised by the factor j_c, which is a measure of the lag on the cooling curve analogous to j for the heating section. Table 3.3 shows the f_H/U against g relationship for different j_c values for z = 10°C (taken from Stumbo [2] who gives more complete tables).

Table 3.3 - f_H/U relationship with g (for z = 10°C)

	Values of $g \times 10^2$ when j_c is:						
f_H/U	0.80	1.00	1.20	1.40	1.60	1.80	2.00
0.30	0.126	0.133	0.141	0.148	0.155	0.163	0.170
0.50	2.63	2.81	2.99	3.17	3.34	3.52	3.69
0.70	9.78	10.5	11.2	12.7	13.4	14.2	
0.90	20.6	22.2	23.8	25.4	27.1	28.7	30.3
1.0	26.9	29.1	31.2	33.3	35.4	37.6	39.7
2.0	85.0	92.2	115	123	130	138	145
3.0	169	181	193	204	216	228	239
4.0	230	245	260	274	289	304	319
5.0	282	300	317	335	353	371	388

j_c is calculated by the same procedure for the cooling curve as used for j in the heating section, i.e. plotting approach to cooling water temperature logarithmically against time.

(e) Having found g from table 3.3 calculate the value of B, the process time, from $B = f_H(\log jI - \log g)$.
Note that this method allows calculation of the process time as an <u>absolute</u> value, since it allows for the D value of the micro-organism. The General Method is <u>relative</u>, in that it only allows comparison of one process lethality with another.

<u>Example 3.2</u>
The product of example 3.1 (processed in a steam retort at 120°C) has an initial number of micro-organisms per can of 1000. This is to be reduced to a probability of survival in each can of 1 in 10^9. Assume a z value of 10°C and a D value of 121.1°C of 1 minute. Calculate the required process time, based on the can centre conditions. The retort come-up-time is 3 minutes.

Following the steps given above:

(a) Figure 3.7 shows the heat penetration curve in log deficit against time form. Note the inversion of the log scale.

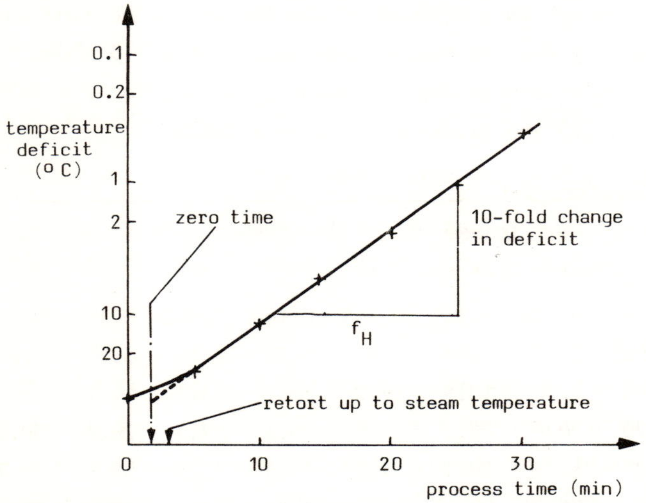

Figure 3.7 - Heat penetration curve for example 3.2

(b) From figure 3.7: f_H = 13.7 min; I = 44.9; p.i.d. = jI = 47; j = 47/44.9 = 1.05. Therefore process time B = 13.7(log 47 - log g)

(c) $F = D[\log a - \log b] = 1\,[\log(1000) - \log(10^{-9})] = 12$

(d) $U = F \times 10^{(121.1 - R.T.)/z} = 12 \times 10^{1.1/10} = 15.46$. Therefore

$f_H/U = 13.7/15.46 = 0.886$. $j_c = 1.23$ from a plot of the cooling curve (figure 3.8). Therefore interpolating for $f_H/U = 0.886$ and $j_c = 1.23$ from Table 3.3, g = 0.23°C.

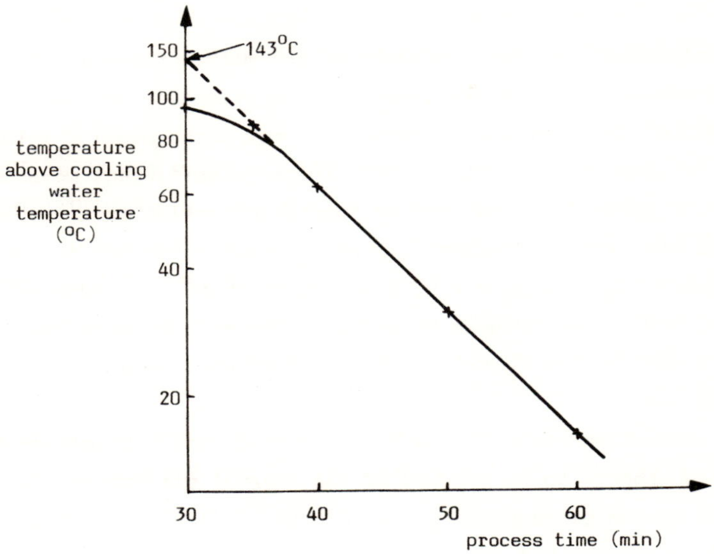

Figure 3.8 - Cooling curve for example 3.2

(e) Therefore B = 13.7 (log 47 - log 0.23) = 13.7 (1.672 + 0.638) = 31.7 min. Thus process time, from steam turned onto the retort, = 31.7 + 0.58x3 = 33.4 minutes.

Example 3.3

In example 3.1 it was calculated that a process of $F_o = 4$ could be obtained with a process time of 21.7 minutes. What probability of survival would this process give for micro-organisms of z = 10°C and (a) D = 240s at 121.1°C (the value for a particularly heat resistant micro-organism) and (b) D = 12s at 121.1°C (the value for Clostridium botulinum)?

(a) B = 21.7 - 0.58x3 = 19.96. Therefore, from $B = f_H(\log jI - \log g)$, 19.96 = 13.7(log 47 - log g); log g = 1.672 - 1.457 = 0.215 and thus g = 1.64°C. Assuming same j_c value (1.23), interpolation from table 3.3 gives $f_H/U = 2.59$. Therefore U = 13.7/2.59 = 5.29. Thus $F = U/10^{(121.1-R.T.)/z} = 5.29/10^{0.11} = 4.11$.

Therefore log a - log b = F/D = 1.03 for D = 4 minutes and 21 for D = 12s. b/a = ratio of surviving to initial micro-organisms $= 10^{-1.03}$ for D = 4 minutes, or 10^{-21} for D = 12s., i.e. just over 1 in 10 for D = 4 minutes and 1 in 10^{21} for D = 12s.

A particularly useful attribute of the mathematical method is that it allows a simple assessment of the effect of change in can size on the sterilsation requirements. For packs heating internally purely by convection, that is where there is sufficient movement of the can contents by natural or forced convection to give no internal temperature gradients, then $f_H \propto$ can volume/surface area, which gives $(f_H)1/(f_H)_2 = r_1h_1(r_2+h_2)/r_2h_2(r_1+h_1)$ for cans of radii r_1 and r_2 and heights h_1 and h_2. Convection packs are characterised by low f_H values (say 5-20 minutes) and j values close to one.

For conduction packs f_H can be calculated from unsteady-state conduction theory to be related to the thermal diffusivity (a) and can size by the expression $f_H = 6.63 \times 10^{-7}/[(1/r^2 + 1.708/h^2)/a]$, where f_H is in minutes, r and h are the can radius and diameter in cm, and a is in m^2/s. Thus $(f_{H1})/(f_{H2}) = [(1/r_2{}^2 + 1.708/h_2{}^2)/(1/r_1{}^2 + 1.708/h_1{}^2)]$. Conduction packs have high f_H values (say 50-200 minutes) and j values around two.

Example 3.4

The process of example 3.2 is carried out with cans of twice the diameter and twice the height of that which gave the heat penetration data of that example. What will the new process time be?

The f_H value in example 3.2 was 13.7 minutes and j = 1.05. The can is therefore a convection pack. f_H for the cans are related by $(f_H)_1/(f_H)_2 = r_1h_1(r_2+h_2)/r_2h_2(r_1+h_1)$. If subscript 2 refers to the original can and subscript 1 refers to the larger can, then $(f_H)_1/(f_H)_2 = 4r_2h_2(r_2+h_2)/r_2h_2(2r_2+2h_2) = 2$. Therefore $(f_H)_1 = 2(f_H)_2 = 27.4$ minutes. Assuming no change in j and using the same steps as in example 3.2: B = 27.4(log 47 - log g); F = 12, U = 15.46 (as before). Therefore $f_H/U = 27.4/15.46 = 1.77$ and g = 0.94°C. Thus, B = 27.4(1.672 - log 0.94) = 27.4 x 1.699 = 46.6

minutes. Therefore process time from steam turned onto the retort $= 46.6 + 0.58 \times 3 =$ 48 minutes. Note that the f_H value for the larger can is somewhat high for convection, but it is still noticeably lower than that for conduction packs.

Example 3.5

(a) What is the f_H value for a can of food material which heats purely by conduction (with a thermal diffusivity of $1 \times 10^{-7} m^2/s$) in a can of 10 cm height and 6 cm diameter? (b) What f_H value would be obtained in a can of this food of double these dimensions? (c) The smaller can takes 80 minutes to reach a temperature 2°C below retort temperature. What temperature will the larger can have achieved in this time?

(a) f_H for smaller can $= 6.63 \times 10^{-7}/[(1/r^2) + (1.708/h^2)a] =$ $6.63 \times 10^{-7}/(0.111 + .0171)(1 \times 10^{-7}) = 51.7$ minutes.

(b) If the dimensions are doubled, r^2 and h^2 are quadrupled and hence f_H is quadrupled. Therefore f_H for larger can $= 51.7 \times 4 =$ 207 minutes.

(c) Since $B = f_H (\log jI - \log g)$; for the smaller can: $80 = 51.7 (\log jI - \log 2)$. Therefore $\log (jI) = 80/51.7 + \log 2 = 1.547 + 0.301 = 1.848$. Thus $jI = 70.47$

Assuming the same value of jI for the larger can: $80 = 207 (1.848 - \log g)$. Therefore $\log g = 1.848 - 80/207 = 1.462$; $g =$ 29.0°C.

Thus the larger can will reach a temperature 29.0°C below retort temperature after 80 minutes.

Other methods of process analysis

A number of alternative procedures are available for analysing other aspects of the canning process. Stumbo [2] has proposed a technique for integrating lethalities over the whole contents of the can to allow for the effect of internal temperature distribution on the total microbiological population of the can. It is also possible to determine the effect of the process on other thermally labile food constituents, e.g. nutrients; Thijssen [4] has recently proposed a procedure for calculating process conditions to give optimum retention of quality characteristics for acceptable sterility. An alternative to in-can sterilisation is aseptic canning, where the food is sterilised outside the can and ascepti-cally filled into sterilised cans. Techniques are available for calculating the sterilisation effects during flow of the food material through the heat exchanger used in this process [5]. A review of the various methods of process analysis is given by Hayakawa [6].

3.2 BULK STERILISATION SYSTEMS

Sterilisation of liquid media for use in the biochemical industries (pharmaceuticals, enzymes etc.) is carried out in both batch and continuous systems, but the principles involved are the same. In order to obtain a reproducible result in processing, it is obvious that the same thermal treatment should be given for all sizes of batch or for all flowrates.

The term 'sterilisation' in these cases means the reduction to an acceptable level of contaminating organisms, usually species which compete for the nutrients with the species being cultivated, or species which produce undesirable side effects, e.g. phage, toxins etc. Thus, in the food industry, it is essential to remove or render inert *Cl. Botulinum*; in most fermentations producing bacterial enzymes *B.Subtilis* is usually the problem.

Sterilisation of media for use in the fermentation industries is usually based on chemical kinetics, assuming a first-order reaction for the death process of spores

$$dN/d\theta = -kN$$

N = number of spores at time θ; k = specific reaction rate for spore destruction.

The specific reaction rate follows the Arrhenius type of correlation with temperature

$$k = A(\exp -E/RT)$$

A = a constant, specific to the species; E = activation energy for spore destruction; R = universal gas constant; T = absolute temperature. Values for A, E and k are published in the liter-ature [7,8,9,10].

Integration of the first-order rate expression at a constant temperature will give the following expression

$$\ln (N_o/N) = k\theta = At(\exp -E/RT)$$

N_o = initial viable spore population.

In food processing, particularly in canning and bottling, F and z values are used. F_{121} is the time required at a temperature of 121°C to achieve an acceptable result in terms of spore destruction, the value of z being the number of degrees of temperature change to reduce the initial spore population by a factor of 10. (F_{121} refers to a pure culture of a particular micro-organism grown on a specific substrate at the specified temperature, whereas F_o (see Section 3.1) refers to the time at 121°C giving the same sterilisation effect as that in an actual canning process assuming z = 10°C).

Thus both F_{121}, z, A and E are specific for a particular species of micro-organism. Similarly the'decimal reduction time' (D) - the time required at a particular temperature to achieve an initial population reduction of a factor of 10 - is also specific to one micro organism.

By definition, $D = 2.303/k$ and $F_{121} = 12D$, since the commonly accepted level of 'commercial sterility' uses $N_o/N = 10^{12}$.

Because all these criteria are inter-related, it is possible to obtain <u>estimates</u> for one type of data from the other type of data.

(a) Having a value for z will allow an estimate of E.

(b) Having a value for D_{121} and z will allow estimation of k_{121} and thus A and E.

(c) Knowing A and E will allow estimation of D_{121} and z (and also a theoretical value for F_{121}).

The data below summarises a variety of information taken from a number of sources.

Data for thermal death rates of micro-organisms.

Species	E (kcal/mol)	A (min^{-1})	D_{121} (min)	F_{121} (min)	z (°C)	Reference
B. Stearothermophilus						
FS 1518	67.7	$10^{37.82}$	1.24*	14.9*	10.2*	7, 11
FS 1518	68.7	$10^{38.13}$	3.0	36.0*	10.0*	8
FS 617	68.7	$10^{38.71}$	0.8	9.6*	10.0*	8
FS 7954	68.7	$10^{38.9}$	0.37	4.45*	10.0*	8
Cl. Sporogenes						
PA 3679	68.7	$10^{38.22}$	1.3	15.6*	10.0*	8
B. Subtilis						
FS 5230	68.7	$10^{37.98}$	0.75	9.0*	10.0*	8
B. Stearothermophilus						
FS 1518	69.24*	$10^{38.53*}$	1.75*	21.0	10.0	12
Cl. Botulinum	69.24*	$10^{39.45*}$	0.21*	2.45	10.0	12
Cl. Sporogenes						
PA 3679	69.24*	$10^{30.07*}$	0.5*	6.0	10.0	12
Thermophile						
PA 3814	99.7*	$10^{56.27*}$	0.25*	3.0	7.0	12

Note: Data asterisked (*) has been derived from the other data.

The values for E derived from z values in the above table are reasonably reliable, but values for A derived from F_{121} measured values, and F_{121} values derived from A and E values should be regarded with some suspicion due to other factors affecting the thermal resistance of micro-organisms.

Batch Sterilisation

Deindoerfer and Humphrey [8] used the symbol ∇ to represent the term $\ln (N_0/N)$, and suggested that ∇ may be regarded as a measure of the success of a sterilisation. In order to determine the treatment given to a batch of material (usually by steam heating in an agitated vessel using jackets or internal coils) the evaluation of ∇ involves a graphical integration over the temperature range of a plot of k values against time. As in food processing, the contribution to sterilisation below a temperature of 100°C is minimal and can be neglected.

Most batch sterilisations involve a heating stage, a holding stage at 121°C and a cooling stage (figure 3.9), and the graphical integration is time consuming.

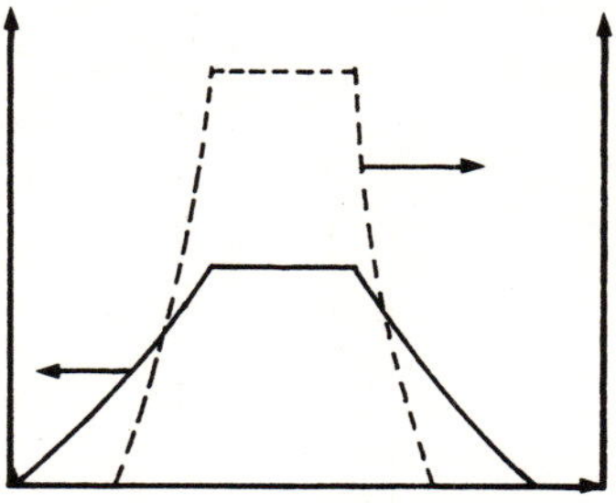

Figure 3.9 - Temperature and k against time.

Richards [11] proposed a rapid calculation method assuming that the temperature profiles above 100°C of the heating and cooling curves were linear.

Assuming that a 'standard' curve shows a temperature increase/ decrease of 1°C/minute, cumulative values of ∇ can be calculated and tabulated for temperatures above 100°C, together with values of k for any organism or species of interest.

Deindoerfer and Humphrey and Richards suggest that a value of $\nabla = 40$ gives adequate spore reduction for most industrial fermentations. The errors introduced by Richards' approximations give a deviation from the rigorous integration of less than 5% (and in most cases 3%). All the workers in this area assume that the contributions from all parts of the sterilisation cycle are additive

$$\nabla_T = \nabla_A + \nabla_H + \nabla_C$$

T = total; A = heating; C = cooling.

It should be noted that when $\nabla_T = 40$, $N_o/N = 2.35 \times 10^{17}$. In canning, $N_o/N = 10^{12}$ for 'commercial' sterility and $\nabla_T = 27.63$. All calculations are based on the statistical probability of spore destruction, and are not absolute.

The most commonly used organism in determining satisfactory sterilisation is B.stearothermophilus 1518, and the following table is based on this organism, where $A = 4.93 \times 10^{37}$ and E =

67480 cal/mol. This data can be used to scale-up from laboratory trials in the absence of data for a particular organism, and is widely used in food processing.

k value and cumulative ∇ values based on 1°C/min temperature rise. B. Stearothermophilus 1518.

Temperature (°C)	k (min^{-1})	∇
100	0.014	-
101	0.018	0.032
102	0.023	0.055
103	0.029	0.084
104	0.037	0.121
105	0.047	0.168
106	0.060	0.228
107	0.076	0.304
108	0.096	0,400
109	0.121	0.521
110	0.153	0.674
111	0.193	0.867
112	0.243	1.11
113	0.306	1.42
114	0.384	1.80
115	0.482	2.28
116	0.604	2.89
117	0.757	3.64
118	0.946	4.59
119	1.182	5.77
120	1.474	7.24
121	1.837	9.18
122	2.286	11.46
123	2.842	14.31
124	3.530	17.84
125	4.380	22.22
126	5.428	27.65
127	6.720	34.37
128	8.310	42.68
129	10.265	52.94
130	12.668	65.61

Example 3.6

The sterilisation cycle for a batch vessel is as follows

heating from 100° to 121° = 28 minutes; holding at 121°C = 30 minutes; cooling from 121° to 100° = 15 minutes. What will be the value of ∇ based on FS 1518 ?

For heating from 100° to 121°C in 21 minutes, the cumulative value from the above table gives ∇ = 9.18.

For heating in 28 minutes, the value of ∇ will be greater because the medium has been held longer at the different temperatures, therefore

$\nabla_A = (9.18 \times 28)/21 = 12.24$

For cooling from 121°C to 100°C in 15 minutes

$\nabla_C = (9.18 \times 15)/21 = 6.56$

Holding at 121°C for 30 minutes, the k value from the above table is 1.837.

$\nabla_H = k\theta = 1.837 \times 30 = 55.11$

Total value for the whole process $\nabla_T = \nabla_A + \nabla_H + \nabla_C$

$\nabla_T = 12.24 + 55.11 + 6.56 = 73.91$

This value is equivalent to $N_o/N = 1.26 \times 10^{32}$.

Example 3.7

Laboratory experiments using 1.0 l vessels showed that the following sterilisation cycle gives adequate results

heating from 100°C to 121°C = 10 minutes

cooling from 121°C to 100°C = 7 minutes

holding time at 121°C = 35 minutes.

These results are to be scaled up to vessels of 5000 l size, and commissioning trials indicate that

time to heat from 100°C to 121°C = 20 minutes

time to cool from 121°C to 100°C = 18 minutes.

How long should the vessel contents be held at 121°C to give the same sterilisation result obtained on the 1.0 l vessels ?

1.0 l vessels.

$\nabla_A = (9.18 \times 10)/21 = 4.37$

$\nabla_C = (9.18 \times 7)/21 = 3.06$

$\nabla_H = 1.837 \times 35 = 64.3$

thus $\nabla_T = 71.73$

5000 l vessels.

$\nabla_A = (9.18 \times 20)/21 = 8.74$

$\nabla_C = (9.18 \times 18)/21 = 7.87$

Since $\nabla_T = 71.73$ to reproduce the conditions using 1.0 l vessels, $\nabla_H = (71.73 - 8.74 - 7.87) = 55.12 = k\theta$.

Thus $\theta = 55.12/1.837 = 30$ minutes.

This rapid method for calculating the effects of thermal treatmen can be used to retrieve a situation where, for example, some malfunction of the control system occurs during a sterilisation cycle.

Example 3.8

During the sterilisation heating cycle of a batch of medium, the control system developed a fault when the temperature reached 115°C. The resulting cycle during heating was

time from 100°C to 115°C = 10 minutes

time at 115°C = 15 minutes.

The normal time to reach 121°C from 100°C is 14 minutes. Assuming that the cooling cycle remains the same as in previous batches, what will be the reduction in holding time at 121°C required to maintain the same total ∇ value as in previous batches ?

Normally, $\nabla_T = \nabla_A + \nabla_C + \nabla_H$. For this abnormal case

$$\nabla_T = \nabla_{A1} + \nabla_{A2} + \nabla_{115} + \nabla'_H + \nabla_C$$

∇_{A1}= rate in heating to 115°C; ∇_{A2} = rate in heating from 115°C to 121°C; ∇_{115} = holding at 115°C; ∇'_H = new holding rate at 121°C.

It can be assumed that all contributions to the sterilisation are additive and $\nabla_A = \nabla_{A1} + \nabla_{A2}$, and ∇_C is the same as a normal batch.

Hence, $(\nabla_H - \nabla'_H) = \nabla_{115} = 0.482 \times 15$ - k value from table.

$= 7.23$

Since k at 121°C = 1.837, the reduction in holding time for this abnormal batch will be 7.23/1.837 $\doteq$ 4.0 minutes.

Continuous Sterilisation

The main aim in continuous sterilisation is to raise the temperature of the medium rapidly, hold for the required time and cool rapidly to operating temperature. Rapid cooling can be achieved by flash evaporation, i.e. by releasing the pressure of the medium after holding and allowing some of the water to flash off as steam. Once the temperature has fallen below 100°C, conventional cooling can be employed.

Example 3.9

A continuous steriliser consists of a plate heat exchanger designed to raise the temperature of the medium to 130°C in 0.5 minute at a flowrate of 0.2 m^3/minute. Assuming that cooling from 130°C to 100°C takes 0.5 minute, how long must the medium be held at 130°C

to achieve a value for ∇ of 74.0 based on FS 1518 ?
From the table of k and ∇ values above

for heating $\nabla_A = (65.61 \times 0.5)/30 = 1.09$

for cooling $\nabla_C = (65.61 \times 0.5)/30 = 1.09$

at 130°C, k = 12.67, and $\nabla_H = k\theta = 12.67\theta$.

$\nabla_T = \nabla_A + \nabla_H + \nabla_C$

so $74.0 = 1.09 + 12.67\theta + 1.09$

$\theta = 5.7$ minutes.

A volume of (0.2 x 5.7) = 1.13 m^3 must be provided for holding.

REFERENCES

1. A.C. Hersom and E.D. Hulland, Canned Foods, 7th Ed., (Churchill Livingstone, Edinburgh, 1980)
2. C.R. Stumbo, Thermobacteriology in Food Processing, 2nd Ed., (Academic Press, New York, 1973)
3. J.G. Brennan et al, Food Engineering Operations, (Elsevier, London, 1969)
4. H.A.C. Thijssen et al, 'Short cut method for the calculation of sterilising times', J.Fd.Sci., 43 (1978) 1096
5. A. Rao and M. Loncin, 'Residence time distribution and its role in continuous pasteurization', Lebensmittel-Wiss. Technol., 7 (1974) 5,14
6. K.I. Hayakawa, 'Mathematical methods for estimating proper thermal processes and their computer implementations', Adv. Fd. Res., 23 (1977) 75
7. F.H. Deindoerfer,'Calculation of heat sterilisation times for fermentation media', Appl. Microbiol., 5 (1957)221
8. F.H. Deindoerfer and A. Humphrey,'Analytical methods for calculating heat sterilisation times', Appl. Microbiol., 7 (1959) 256
9. S. Aiba, A.E. Humphrey and N.F. Mills, Biochemical Engineering, (University of Tokyo Press, 1973)
10. N. Blakebrough, Biochemical and Biological Engineering, Vol.2, (Academic Press, London, 1968)

11. J.W. Richards, 'Rapid calculations for heat sterilisations', <u>Br. chem. Engng</u>, 10 (1965) 166
12. S.E. Charm, <u>Fundamentals of Food Engineering</u>, 2nd Ed., (AVI, Westport, 1971)

4 HEAT AND MASS TRANSFER

4.1 DRYING

There are three major industrial methods of removing moisture from solid materials:
(a) by subjecting it to a high velocity stream of heated, low humidity air (air drying) or (b) by placing it on a heated surface and allowing evaporation of moisture into the surrounding atmosphere (contact drying) or (c) by subjecting it to a low pressure and a heating source (vacuum drying). A variation on the latter process is freeze-drying, where the product is initially frozen, placed in a very low pressure environment (such that the water is below its triple point and it exists only in vapour or solid form) and then heated to sublime off the moisture. Keey [1] and Van Arsdel [2] describe the science and technology of these processes. The major problem in process design in drying operations is to estimate the time to achieve the desired final moisture content.

Air drying

Figure 4.1 shows typical drying characteristics of wet solids in air.
Ignoring the very early stage during which the product is heating up there are two major stages of drying: (a) a first period in which moisture is comparatively easy to remove and (b) a second period in which moisture is bound, or held, within the solid matrix. The moisture content at which the change from the first to second period occurs is known as the critical moisture content.

First period. If heat transfer is solely by convection from the drying air, the surface temperature drops to the wet-bulb temperature of the air as latent heat of evaporation is taken up by the film of water at the surface. The wet bulb temperature can be determined from psychrometric data for moist air as shown in figure 4.2.

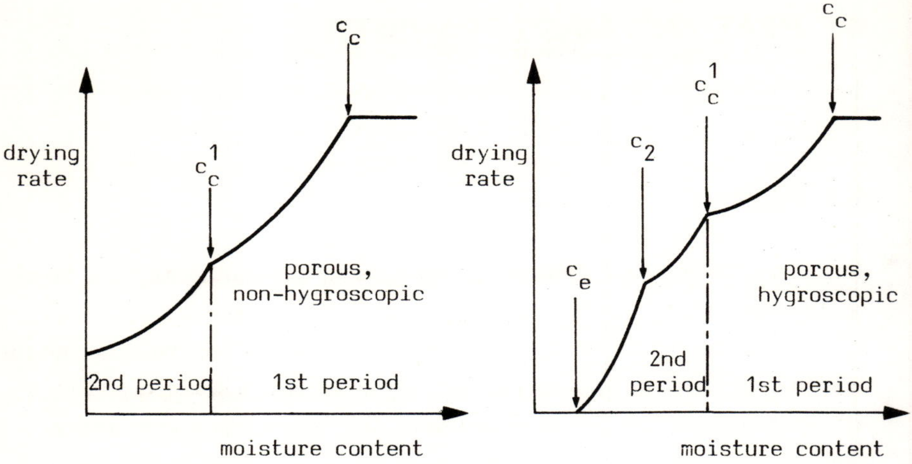

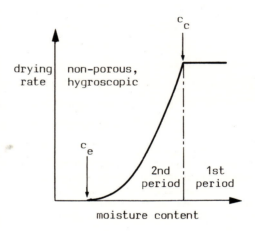

Figure 4.1 - Drying characteristics of wet solids

For as long as free water exists at the surface the drying rate is constant and given by $F_1 = h(t_a - t_s)/\lambda$, where F_1 is the rate of moisture loss per unit area per unit time, λ is the latent heat of evaporation and t_a and t_s are the air and surface temperatures respectively. For solid materials in slab form on trays or belts in a drier the air flow is commonly either parallel to, or perpendicular to, the slab surface. The heat transfer coefficient in these two cases can be calculated [4] from either $h = 14.3\ G^{0.8}$

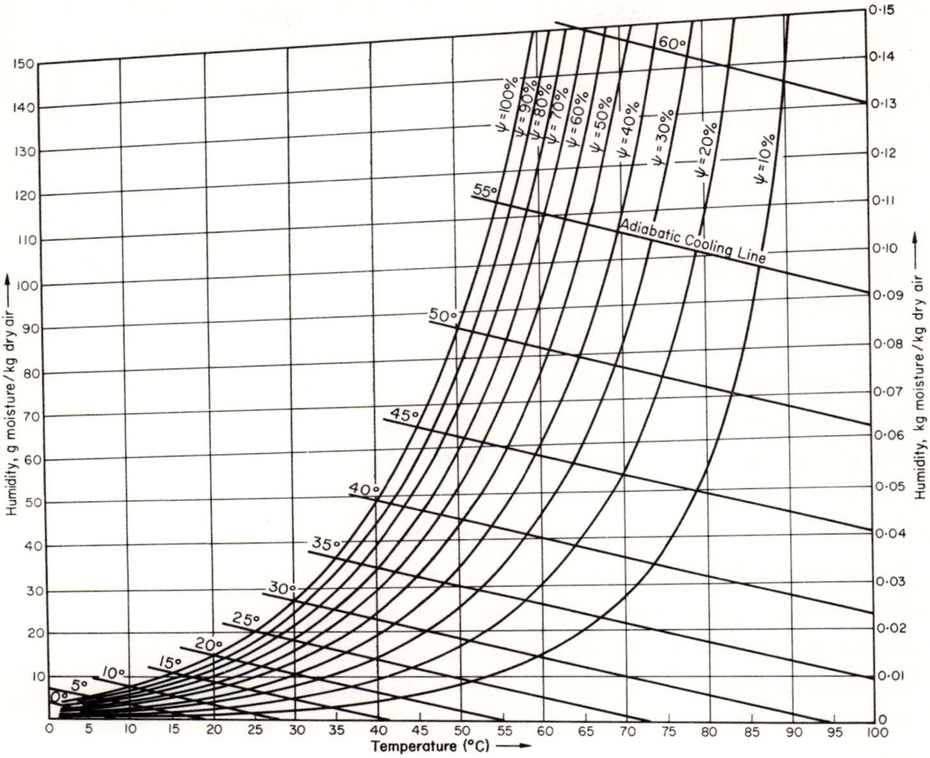

To obtain the wet bulb temperature, locate the adiabatic cooling line which passes through the intersection of the lines representing the air relative humidity (ψ) and the dry bulb air temperature (x-axis). This line passes through ψ = 100% at the wet bulb temperature.

Figure 4.2 - Psychrometric chart for air at atmospheric pressure (from [3])

(parallel flow) or $h = 24.2\ G^{0.37}$ (perpendicular flow), where G is the air mass velocity in kg/m^2s and h has units of W/m^2K. The air mass velocity = mass flow rate/cross-sectional area of flow = volumetric flow rate x air density/cross-sectional area = linear air velocity x air density.

For <u>non-porous materials</u> the rate remains constant until the critical moisture content (c_c) is reached. Marshall and Friedman [5] reported figures for the critical moisture content of a number of materials ranging from 3% to 450% (on a dry weight basis) with the great majority below 50%. For foodstuffs values are generally high, e.g. 240%-770% for a variety of fruits and vegetables [6], 300-450% for fish muscle [12], 400% for gelatin [5], 300% for peas [7].

Example 4.1

Ten pieces of leather, each weighing 2.5kg and having a surface area of $1m^2$, are pasted on glass plates and hung in a tunnel air drier of free cross-sectional area $2m^2$. Air at 30% R.H. and 60°C is blown over the leather, parallel to their surfaces, at a flow rate of 8 m^3/s. The initial moisture content of the leather is 120% (dry weight basis). How long will it take to dry the leather to its critical moisture content of 95% ? (air density = 1.1 kg/m^3 and latent heat of evaporation = 2300 kJ/kg)

$G = 8 \times 1.1/2 = 4.4\ kg/m^2s$; $h = 14.3\ (4.4)^{0.8} = 46.8\ W/m^2K$. Air at 60°C, 30% R.H. has a wet-bulb temperature of 42°C (from figure 4.2). Therefore $t_a - t_s = 60-42 = 18$°C and $F_1 = 46.8 \times 18/2300 \times 10^3 = 3.66 \times 10^{-4}\ kg/m^2s$. Therefore drying rate for all ten pieces = 3.66×10^{-3} kg/s. Initially leather contains 25 x (120/220) = 13.64 kg moisture and 11.36 kg dry matter. At c_c the leather contains 11.36 x (95/100) = 10.79 kg moisture. Therefore water to be removed = 2.85 kg. Therefore time to c_c = $2.85/3.66 \times 10^{-3}$ = 779s = 13.0min.

If the drying material is porous the rate of evaporation in this first stage of drying becomes lower than that predicted for a free-water surface and falls as a second critical moisture content (c_c^1) is approached. The first period of drying ends when water held in the funicular state is removed [1]. A reasonable approximation to the moisture content at the end of this period of drying is given by $v_c/v_i = 0.2$, where v_i and v_c are the mean volumetric water contents at the beginning and end of this period. The average rate of moisture loss during the first period of drying of a porous body is difficult to predict, and is best established from experimental evidence. In their absence, a possible estimate would be 40% below the free water rate of evaporation. If the material being dried is also hygroscopic (that is capable of retaining moisture in a bound form within its structure, the amount of which is dependent on the surrounding atmospheric relative humidity), there may be a more complicated first stage of drying. In this case the first stage of drying can be considered completed when all moisture in an unbound state has been removed from the structure. In the absence of experimental data describing the drying characteristics of such a material this can be taken to occur at a moisture content $(c_c^1) = 0.4c_c$.

Example 4.2

Spherical nylon beads which have been soaked in water are dried by placing them in a 40 mm thick layer on trays of total surface area $4m^2$ over which air is blown at 5 m/s. The bed has a voidage (fractional free space) of 0.40 and an initial moisture content of 70%; the bead density is 1100 kg/m^3. The air is at 85°C and 15% relative humidity. How long will the first period of drying be? (Air density = 1.0 kg/m^3, latent heat = 2300 kJ/kg)

By the same procedure as example 4.1; G = 5 kg/m^2s, h = 51.8 W/m^2K, t_s = 57°C, $F_1 = 6.31 \times 10^{-4}$ kg/m^2s. Bed volume = $0.16m^3$, bead volume = 0.6 x 0.16 = $0.096m^3$, bead mass = 0.096 x 1100 = 105.6 kg. Thus if voids are completely full of water, water retained in pores = 0.064 x 1000 = 64 kg = (64/105.6) x 100% = 60.6%. Since the bed has an initial moisture content of 70% free water can be evaporated at a rate F_1 until the moisture content drops to 60.6%. Water to be removed in this state = (0.70 x 105.6) - 64 = 9.92 kg. Therefore time for this stage = $9.92/6.31 \times 10^{-4}$ x 4 = 3930s = 65.5min.

Volumetric water content ratio at end of first period can be taken as 0.2 x 0.40 = 0.08. Mass of water retained at this point = 0.08 x 0.16 x 1000 = 14.4 kg. Thus water to be removed in the slower stage of the first period = 64 - 14.4 = 49.6 kg. Assume an average rate of 40% below free rate, i.e. 3.79×10^{-4} kg/m^2s, giving a time in this stage of $49.6/3.79 \times 10^{-4} \times 4 = 3.28 \times 10^4 s$ = 546 min. Therefore total time = 66 + 546 = 612min.

Second period. The removal of moisture in this period takes place by complex phenomena. If the material is porous and non-hygroscopic Keey [1] has proposed that the rate of drying is given by $F_2 = (F_c - F_e)(c/c_c^1)^2 + F_e$, where F_c, F_e and F_2 are the drying rates at the end of the first period, at the end of the second period and at an average moisture content of c, respectively. By integration over the whole period for a slab this equation gives a drying time in this period $\theta_2 = L\rho c_c^1[-\tan^{-1}(F_c^*/c_c^1) + \tan^{-1} F^*]/[(F_c-F_e)F_e]^{0.5}$, where $F^* = [F_c-F_e)/F_e]^{0.5}$, L is the material thickness for one-sided drying or half the material thickness for two-sided drying, ρ is the dry bulk density. F_e is obtained from [1]: $F_e = F_1/(Bi + 1)$, where Bi is the Biot Number = [diffusion resistance coefficient (μ_D) x material thickness x heat transfer coefficient]/[air density x air humid heat x diffusion coefficient for water vapour in still air (D_{AB})]. μ_D has been measured for a number of materials [1], typically taking a value in the range 1.5 to 7. In the absence of experimental evidence, a reasonable figure for F_c is $0.4F_1$.

Example 4.3

The nylon beads of example 4.2 are to be dried to a final moisture content of 8%. How long will this take ? (Air humid heat = 1300 J/kgK, $D_{AB} = 3.6 \times 10^{-5}$ m^2/s)

Kessler [8] gives $\mu_D = 4.12$ for randomly packed spherical particles. Thus $Bi = 4.12 \times 0.04 \times 51.8/1.0 \times 1300 \times 3.6 \times 10^{-5} = 182$. $F_e = F_1/Bi + 1 = 6.31 \times 10^{-4}/183 = 3.45 \times 10^{-6}$ kg/m^{2}s. $F_c = 0.4 \times 6.31 \times 10^{-4} = 2.52 \times 10^{-4}$ kg/m^{2}s. $F^* = [(252-3.5)/3.45]^{0.5} = 8.49$. $c_c^1 = 14.4/105.6 = 0.136$ (see example 4.2). Therefore $\theta_2 = [0.04 \times (105.6/0.16) \times 0.136] [\tan^{-1}8.49 - \tan^{-1}(8.49 \times 0.08/0.136)]/(248 \times 10^{-6} \times 3.45 \times 10^{-6})^{0.5} = 9872s = 164min$.

For a porous hygroscopic material there is often a break in the second period as moisture bound by different mechanisms to the structure is progressively removed. Up to this break it has been proposed [1] that the rate of drying (F_2) at a concentration c is given by $F_2 = (F_c-F_e)(c/c_c^1)^n + F_e$, where n is approximately 0.5

for fibrous materials and 1 for other hygroscopic substances. The following equations for drying time in this stage (i.e. to reach a moisture content c_2 at the break point) can be derived for these two cases: $\theta_2 = [2L\rho c_c^{1\,0.5} F^{**}/F_e]\{(c_c^{1\,0.5} - c_2^{0.5}) - F^{**} c_c^{1\,0.5} \ln[(c_c^{1\,0.5} + F^{**} c_c^{1\,0.5})/(c_2^{0.5} + F^{**} c_c^{1\,0.5})]\}$ for fibrous materials, where $F^{**} = F_e/(F_c - F_e)$, and $\theta_2 = [L\rho c_c^1/(F_c - F_e)]\ln\{F_c c_c^1/[(F_c - F_e)c_2 + F_e c_c^1]\}$ for other hygroscopic materials. In the absence of experimental drying data, c_2 can be taken as $c_c^1/3$.

After the break in the second period the rate of drying falls approximately linearly with moisture content until the material reaches its equilibrium moisture content (c_e) i.e. the moisture content of a hygroscopic substance when in equilibrium with a surrounding atmosphere of constant relative humidity. Values of c_e are known for a number of foods [9]. A linear fall gives a time θ_3 to achieve a final moisture content of c_3, where $\theta_3 = [L\rho(c_2 - c_e)/F_3]\ln[(c_2 - c_e)/(c_3 - c_e)]$ and F_3 is the drying rate at the break point.

Example 4.4

30kg of potato granules with an initial moisture content of 60% (dry weight basis) are to be dried by placing them on a perforated belt of $5m^2$ surface area and blowing air perpendicularly over their surfaces. The bed is 15mm deep. The air is at 75°C, 20% R.H. and has a mass velocity of $5kg/m^2s$. What time will be required to dry the granules to an average moisture content of 8% if the equilibrium moisture content and critical moisture content (c_c) are 3% and 120% respectively? (Dry bulk density = $300kg/m^3$; air density = $1.0kg/m^3$; latent heat = 2300kJ/kg; humid heat = 1300J/kgK; $D_{AB} = 3.6 \times 10^{-5} m^2/s$)

This material can be considered a porous, hygroscopic material. First period: since the initial moisture content is below c_c there is no constant rate drying stage. In the absence of further information assume $c_c^1 = 0.4c_c = 0.48$ and the average drying rate to c_c^1 is $0.6F_1$ (see above). $h = 24.2 \times 5^{0.37} = 43.9W/m^2K$, $t_s = 50°C$, giving $F_1 = 4.77 \times 10^{-4}kg/m^2s$. Thus evaporation rate = $2.86 \times 10^{-4}kg/m^2s$. Initial moisture content = (60/160) x 30 = 11.25kg and at c_c^1 = 0.48 x18.75 = 9kg. Thus time for the first period is $2.25/2.86 \times 10^{-4} \times 10$ = 787s = 13.1 min.

Second period: Krischer [10] gives $\mu_D \doteq 2$, therefore Bi = 2 x 0.0075 x 43.9/(1 x 1300 x 3.6 x 10^{-5}) = 14.1 (using 0.5L in Bi since internal diffusion is only through half thickness of slab). F_e = 4.77 x 10^{-4}/ 15.1 = 3.16 x 10^{-5}kg/m^{2}s. Take $c_2 = c_c^1/3$ = 0.16 and $F_c = 0.4F_1$ = 1.91 x 10^{-4}kg/m^{2}s. Then θ_2 = [0.0075 x 300 x 0.48/1.59 x 10^{-4}]ln[1.91 x 10^{-4} x 0.48/(1.59 x 10^{-4} x 0.16 + 3.16 x 10^{-5} x 0.48) = 5531s = 92.2min. $F_3 = F_2$ at c_2 = 1.59 x 10^{-4} x (1/3) + 3.16 x 10^{-5} = 8.46 x 10^{-5}kg/m^{2}s. Thus θ_3 = [0.0075 x 300 x 0.13/8.46 x 10^{-5}]ln[0.13/ 0.05] = 3304s = 55.1min. Therefore total drying time = 160min.

For hygroscopic non-porous materials such as gels, the drying time in the second period to reach a moisture content of c is best described by the equation $\ln[(c-c_e)/c_c-c_e)] = \ln(8/\pi^2)-(\pi^2 D\theta)/4L^2$ for a slab drying from one side (L is the half-thickness for a slab drying from two sides). For a sphere the equivalent expression is $\ln[(c-c_e)/(c_c-c_e)] = \ln(6/\pi^2)-(\pi^2 D\theta)/4L^2$ where L is the sphere diameter. Both expressions are the first terms of the series solutions equivalent to that for heat transfer given in Chapter 2.2. D is the diffusion coefficient for liquid water in the substance and is assumed constant. If there are no changes in the manner of the binding of moisture to the matrix this is a reasonable assumption (e.g. Bakker-Arkema[11] found a constant D in drying alfalfa wafers). Normally, however, D falls as the moisture content falls. Although it is possible to establish analytical solutions for the diffusion equations for some known D:c relationships there is normally insufficient information on particular materials to make these useful. A possible approximation is to assume a step change in D at some particular moisture content, e.g. Jason [12] found a five-fold change in D at a moisture content of 15% fitted experimental results on fish drying. Using this approach, the diffusion equation above can be used to determine the drying time up to the step change in D and, assuming a parabolic distribution at that time, the second part of the drying will be approximated by: $[(c-c_e)/(c_b-c_e)] = [2.4/(c_e + 2c_{bc})]\exp[-D\pi^2\theta/4L^2][0.8c_{bc} + 0.2c_e]$ where c_b and c_{bc} are respectively the average and centre moisture content of a slab at the start of the second stage. c_{bc} can be calculated from $(c_{bc}-c_e)/(c_c-c_e)=(4/\pi)\exp[-D\pi^2\theta/4L^2]$, using D for the first stage.

Example 4.5

Apple slices, 6cm diameter and 3mm thick, are dried in an air stream with air blowing horizontally over their major surfaces. Their initial moisture content is 80% (dry weight basis) and they are to be dried to 5% moisture content. Their critical moisture content is 250% and their equilibrium moisture content (at the temperature and R.H. of the drying air) is 3%. Calculate the drying time assuming (a) a constant diffusion coefficient of $5 \times 10^{-9} m^2/s$ and (b) a two-stage process with an initial D of $1 \times 10^{-8} m^2/s$ falling to $D = 2 \times 10^{-9} m^2/s$ at a moisture content of 20%.

There is no constant rate period since c_c is higher than the initial moisture content.

(a) $\ln[(0.05-0.03)/(0.80-0.03)] = \ln(8/\pi^2) - (\pi^2 \times 5 \times 10^{-9} \times \theta)/ 4 \times 0.0015^2$, which gives $\theta = 630s$.

(b) For the initial period to 20% moisture content, by the same procedure as in (a), $\theta = 119s$. For the second period, $(c_{bc}-0.03)/(0.8-0.03) = (4/\pi)\exp[-1 \times 10^{-8} \times \pi^2 \times 119/4 \times 0.0015\ ]$, which gives $c_{bc} = 0.296$. Therefore $[(0.05-0.03)/90.2-0.03)] = [2.4/(0.03 + 0.592)]\exp[- 2 \times 10^{-9} \times \pi^2 \times \theta/4 \times 0.0015^2][0.8 \times 0.296) + (0.2 \times 0.03)]$, which gives $\theta = 946s$. Thus the total time is 1065s.

The drying of particulate material in a fluidised bed is discussed in Chapter 7. The above procedures allow estimates of drying time assuming minimal experimentation to determine drying characteristics. It will be seen that such experiments are essential for accurate assessment of drying rates. Thijssen [13] has suggested procedures for establishing rates from simple drying tests on product samples. Keey [1] gives procedures for predicting drying times when the drying air changes in moisture content as it passes through the drier.

Contact drying

When a moist material is placed on a heated surface, heat flows by conduction to the free surface of the material. Some of this heat causes evaporation to occur, some is lost by convection from the surface and some causes the surface of the material to rise in temperature. While there is still sufficient moisture to keep the surface moist an equilibrium temperature will be achieved. If the temperature of the air, product surface and heated surface are t_a, t_s and t_p respectively, a heat balance gives $k_m(t_p-t_s)/L = h(t_s-t_a) + h(H_s-H_a)\lambda/S$, where k_m is the thermal conductivity of the moist product, H_s and H_a are the absolute humidities of the air at the surface and in the bulk air stream and S is the humid heat of the air.

As drying proceeds below the critical moisture content the drying rate falls because non-superficial water is harder to remove (as in air drying). When the finalquantities of moisture are being removed from a porous material (or up to equilibrium in hygroscopic materials), the rate of drying (F_e) will either be controlled by diffusion of moisture vapour from the hot surface through the nearly dry porous structure or by the potential evaporation rate at the surface, whichever is the smaller. The latter will be more probable at low contact temperature. For diffusion control, $F_e = h(H_p-H_a)/S(1+Bi)$, where H_p is the absolute humidity at the heating plate temperature. For surface evaporation control, $F_e = h(H_s-H_a)/S$. Having calculated F_e drying times for the whole process can then be obtained as for air drying.

Example 4.6

100kg of a fibrous material, in the form of a mat 20mm thick and dry bulk density 300kg/m^3, is dried on a heated plate (temperature 110°C) and air is circulating over the surface at $G = 0.5$kg/m^2s, temperature 30°C,30%R.H.The material is initially at 45% moisture content (dry weight basis), its critical moisture content is 30% and equilibrium moisture content is 3%. The density of the dry solids are 1600kg/m^3 and the thermal conductivities of the moist and dry material are 0.25W/m K and 0.05W/m K. Take S = 1300J/kg K and λ = 2300kJ/kg. Calculate the time to dry the mat to 5% moisture content. Assume a Biot Number of 35.

Initial drying rate (F_1) is obtained from the heat balance equation. H_s depends on t_s and therefore t_s has to be obtained by trial and error. For t_s = 40°C, H_s = 0.049kg/kg (from psychrometric chart).

H_a = 0.008kg/kg. Thus $k_m(t_p-t_s)/L$ = 0.25 x 70/0.02 = 875 ; $h(t_s-t_a)$ = 14.3 x $0.5^{0.8}$ x 10 = 82 ; $h(H_s-H_a)\lambda/S$ = 14.3 x $0.5^{0.8}$ x 0.041 x 2.3 x 10^6/1300 = 596 . Therefore conduction term is too large. The same calculation for t_s = 50°C gives too large an evaporation term. By iteration, the correct value for t_s = 43°C and $F_1 = h(H_s-H_a)/S$ = 3.16 x 10^{-4} kg/m^{2}s. Initial moisture content = 100 x (45/145) = 31.0 kg. Moisture content at c_c = 0.30 x 69.0 = 20.7kg. Therefore 10.3kg moisture is removed in this stage. Surface area of mat = total volume/thickness = (68.96/300)/0.02 = 11.5m^2. Thus drying time in this stage = 10.3/3.16 x 10^{-4} x 11.5 = 2834s = 47.2min. In the second stage, possible final evaporation rate from the surface is calculated as for the first stage (using k for the dried material), giving a surface temperature of approximately 28°C and F_e = 1.00 x 10^{-4}kg/m^{2}s. At 110°C the potential vapour production rate at the hot surface is effectively infinite (because of the high temperature) and thus internal diffusion is not limiting. Therefore surface evaporation controls and F_e = 1.00 x 10^{-4}kg/m^{2}s. (If the surface temperature was 80°C this calculation shows that diffusion controls and the rate is 0.97 x 10^{-5}kg/m^{2}s).

Since the material is hygroscopic and porous, there will be a two part second stage of drying (see example 4.4). Using the equation for θ_2 for a fibrous material, F^{**} = 1 x 10^{-4}/(3.16 x 10^{-4}-1 x 10^{-4}) = 0.463 and using c_2 = 0.30/3 = 0.10, θ_2 = [2 x 0.02 x 300 x $0.3^{0.5}$ x 0.463/1 x 10^{-4}]{($0.3^{0.5}$ - $0.1^{0.5}$) - 0.463 x $0.3^{0.5}$ln [($0.3^{0.5}$ + 0.463 x $0.3^{0.5}$)/($0.1^{0.5}$ + 0.463 x $0.3^{0.5}$)] = 4414s = 73.6 min. θ_3 is calculated as in example 4.4, giving 2844s = 47.4 min. Therefore total time = 168 min.

Contact drying of non-porous materials is identical in analysis to air drying, except that the diffusion coefficient is higher (since it increases with temperature), which reduces the drying time. In the absence of information on the temperature dependence of D for a particular material, a value of 40kJ/g mol is a typical figure for the activation energy E in the relationship D_{t1}/D_{t2} = $\exp[E/Rt_2-E/Rt_1]$, where D_{t1} and D_{t2} are the diffusion coefficients at temperatures t_1 and t_2 respectively, and R is the universal gas constant.

Freeze drying

In freeze drying the material which has been frozen is subjected to a pressure below the triple point value (630 Pa) and heated to cause ice sublimation to vapour. This results in very little damage being caused to the product because of the low temperature employed and the absence of diffusing liquid water. It is therefore used for easily degraded materials such as foodstuffs.

Heat for sublimation can be transmitted either through the dry porous layer produced as the ice front recedes inwards from the surface or through the remaining frozen layer (Figure 4.3).

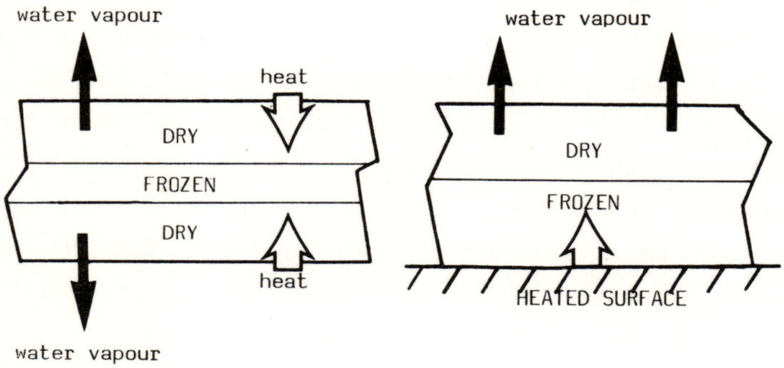

Figure 4.3 - Models for freeze drying analysis

Heating through dry layer. The following equation has been proposed [14] for evaluating the drying time: $\theta = L^2\rho(c_o-c_f)\lambda_s/2k_d(t_s-t_i)$ for a slab of half-thickness L drying from two sides, where c_o and c_f are the initial and final moisture contents, λ_s is the latent heat of sublimation, k_d is the thermal conductivity of the dry layer, t_s and t_i are the surface and ice-front temperatures. A heat and mass balance at the ice-front shows that t_i will remain constant and is related to the pressure at the ice-front (p_i) by $p_i = p_s + (k_d/b\lambda_s)(t_s-t_i)$ where b is the permeability of the dry porous layer and p_s is the pressure in the vacuum chamber. Since p_i also has a known thermodynamic relationship to t_i, both p_i and t_i can be calculated.

Example 4.7

Slices of a food product 5mm thick are freeze-dried with both major surfaces of the slices kept at 50°C in the vacuum chamber. The chamber pressure is 60Pa. Calculate the time to dry from 200% moisture content (dry weight basis) to 5%. (k_d=0.02W/mK, b= 2 x 10^{-8} kg/s m Pa, λ_s = 2.95 x 10^6J/kg, ρ = 500kg/m^3)

p_i = 60 + (0.02/2 x 10^{-8} x 2.95 x 10^6)(50-t_i) = 76.9 - 0.339t_i. Figure 4.4 shows this equation plotted on the same graph as the thermodynamic pressure-temperature relationship for ice. Hence p_i and t_i are 85 Pa and -22°C respectively. Therefore θ = $(0.0025)^2$ x 500 x (2-0.05) x 2.95 x 10^6/2 x 0.02 x (50+22) = 6242s = 104 min.

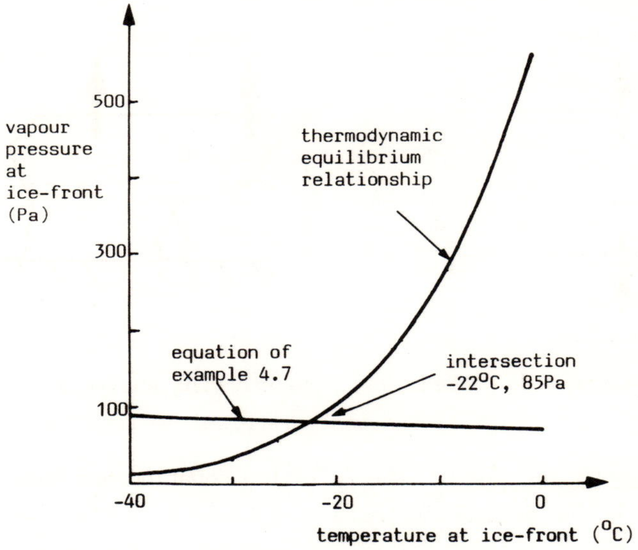

Figure 4.4 - Pressure:Temperature relationships at ice interface in freeze drying

Heating through frozen layer. In this case p_i is not independent of the dried layer thickness (x_d). If the frozen materials rests on a surface heated by a heat transfer fluid at temperature t_h then $p_i = p_s + (k_f/\lambda_s b)(hx_d/[k_f+hx_f])(t_h-t_i)$, where k_f is the thermal conductivity of the frozen layer and h is the heat transfer coefficient from the heating fluid to the frozen material surface. p_i has to be calculated for various values of x_d as the frozen layer retreats into the material and the drying time is then given by $\theta = [\int_0^{2L} x_d dx_d/(p_i-p_s)]\rho(c_o-c_f)/b$.

Example 4.8

The food product of example 4.8 is freeze-dried in the same vacuum chamber, but with the slices resting on a surface heated by a heat transfer fluid at 60°C. Calculate the drying time if k_f = 1.2W/m K and h = 100W/m^2K.

p_i = 60 + (1.2/2.95 x 10^6 x 2 x 10^{-8}) 100x_d/[1.2 + 100 (0.005 - x_d)] (60 - t_i). Substituting values of x_d from 0 to 5 x 10^{-3}m in this equation gives the p_i:t_i relationship as in Figure 4.4. The points of intersection with the thermodynamic relationships give the following values:

x_d(m)	p_i(Pa)	p_i-p_s(Pa)	x_d/[p_i-p_s](m/Pa)	t_i(°C)
0	60	0	-	-25.5
5x10^{-6}	60.52	0.52	9.62x10^{-6}	-25.3
5x10^{-5}	65.07	5.07	9.86x10^{-6}	-24.5
5x10^{-4}	109	49	10.20x10^{-6}	-19.5
1x10^{-3}	156	96	10.42x10^{-6}	-15.6
2x10^{-3}	250	190	10.53x10^{-6}	-10.2
3x10^{-3}	350	290	10.34x10^{-6}	-6.7
4x10^{-3}	456	396	10.10x10^{-6}	-3.5
5x10^{-3}	575	515	9.71x10^{-6}	-0.8

Since x_d/(p_i-p_s) is fairly constant graphical integration is not necessary and an average value is taken for $\int_0^{0.005} x_d\, dx_d/(p_i-p_s)$ = 10.25 x 10^{-6} x 0.005 = 5.13 x $10^{-8}$$m^2$/Pa. Hence θ = 5.13 x 10^{-8} x 500 x 1.95/2 x 10^{-8} = 2501s = 42min.

Note that the ice surface reaches approximately the melting point at the end of the process. This is a problem associated with heating through the frozen layer and the above calculation suggests a lower heating temperature should be used.

Heating by radiation to the drying surface. When heating through the dried layer it was assumed in the above procedure that there was negligible thermal resistance at the surface. Allowing for the low thermal conductivity of the dry layer this is a reasonable assumption in many cases. However, a finite heat transfer can be allowed for in the analysis.

A heat balance with a finite heat transfer coefficient leads to the equation $p_i = p_s + (k_d/b\lambda_s)(hx_d/k_d+hx_d)(t_s-t_i)$ and the drying time can be calculated by the method given above for the case of heat transfer through the frozen layer.

Karel[14] gives typical values for the properties of foodstuffs used in the above calculations.

4.2 BOILING HEAT TRANSFER AND PEAK HEAT FLUX

If a heated element is immersed in a liquid, the transfer of heat and the pattern of boiling as the 'excess' temperature (ΔT_x) increases, will be of the form shown in figure 4.5. The excess temperature is defined as

$$\Delta T_x = (t_s - t_1)$$

t_s = heated surface temperature; t_1 = saturation temperature of the liquid.

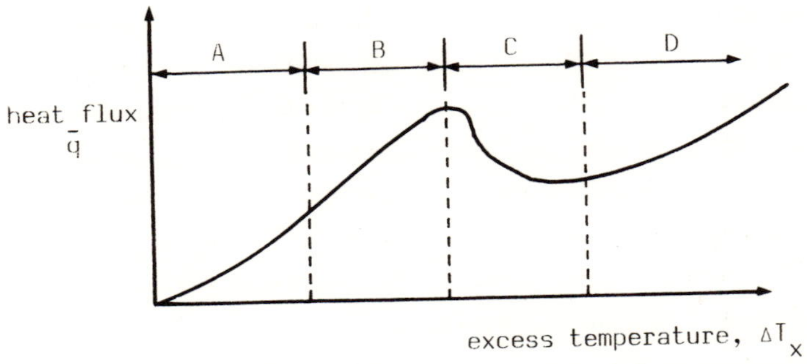

Figure 4.5 - Heat flux versus excess temperature

The regions shown on figure 4.5 are as follows

Interface evaporation. (A) Entirely natural convection, the heat being transferred by superheated liquid rising to the surface where evaporation takes place.

Nucleate boiling. (B) Bubbles of vapour form at numerous sites of nucleation on the heated surface and exert considerable agitation on the liquid as they rise to the surface. As the excess temperature increases, the bubbles become continuous rising streams of vapour. This region of boiling is analogous to forced convection of the liquid at the heated surface.

Unstable film boiling. (C) At some critical excess temperature, the bubbles of vapour generated at the surface coalesce and form a blanket of vapour over the surface. The heat flux rapidly decreases once this stage is reached.

Stable film boiling. (D) As the excess temperature is further increased, the vapour film generated in region C becomes stable, and radiation from the heated surface starts to play an increasing role in the heat transfer.

The most important region for process operation is in the nucleate boiling zone. The heat flux ($\bar{q}$), and hence the heat transfer coefficient (h_b), depend on the properties of the surface as well as the properties of the liquid being heated, and as well as the prediction of h_b, the prediction of the peak heat flux at the point where the boiling mechanism changes to the unstable film region is of importance in design.

This type of boiling where no mechanical means of agitation is employed is known as 'pool boiling', and a number of correlations exist for the determination of $\bar{q}$ and h_b, as well as for the prediction of the critical heat flux. These correlations can be applied, with modifications, to boiling combined with forced convection.

In all cases of pool boiling, the heat transfer relationship is defined by

$$\bar{q} = h_b \Delta T_x \quad \text{where } \bar{q} = q/A.$$

Prediction of Heat Transfer Coefficients

Rohsenow [15] proposed the following relationship for the variation of heat flux with excess temperature

$$(C_p \Delta T_x/\lambda) = K[(\bar{q}/\mu\lambda)\sqrt{(\psi)}]^{1/3} \ (Pr)_L^{1.7}$$

and $\psi = (\sigma/g\Delta\rho)$

C_p = specific heat of the liquid; ΔT_x = excess temperature; λ = latent heat of vaporisation of the liquid; $\bar{q}$ = heat flux; μ = liquid viscosity; σ = surface tension of the liquid; g = acceleration due to gravity; $\Delta\rho = (\rho_L - \rho_V)$; ρ_L = liquid density; ρ_V = vapour density; $(Pr)_L$ = Prandtl number for the liquid; K = Rohsenow constant.

The constant (K) in the Rohsenow correlation is dependent on the particular liquid/surface combination, and in order to apply this correlation for all cases of the same liquid and surface, one experimental value of $\bar{q}$ and ΔT_x must be known, and K for the combinatior can be evaluated.

Vachon et al [16] have investigated this correlation for different surfaces, and selected values are given below.

Values of the Rohsenow constant

liquid/surface combination	K
water/scored copper	0.0068
water/emery polished copper	0.0128
water/ground, polished SS	0.008
water/chemically etched SS	0.013
water/mechanically polished SS	0.0132

Kutateladze [17] proposed a similar correlation

$$(h_b/k)\psi^{1/2} = 0.0007\ [(\bar{q}/a\lambda\rho_V)(P/\sigma)\psi]^{0.7}\ (Pr)_L^{-0.35}$$

a = thermal diffusivity of the liquid = $(k/C_p\rho_L)$; k = thermal conductivity of the liquid; P = absolute pressure.

An empirical relationship applicable to boiling both inside and outside of tubes is that due to Gilmour [18]

$$(h_b/C_pG)(Pr)_L^{0.6}(\rho_L\sigma g/P^2)^{0.425} = 0.001/(DG/\mu)^{0.3}$$

D = diameter of the tube; $G = (V\rho_L/A\rho_V) = (\bar{q}\rho_L/\lambda\rho_V)$; V = mass vapour flowrate.

According to Palen and Taborek [19] who tested 14 correlations for the boiling heat transfer coefficient, the most reproducible was that due to McNelly [20], which is a semiempirical relationship

$$(h_bD/k) = 0.225(Pr)_L^{0.69}(\bar{q}D/\lambda\mu)^{0.69}(PD/\sigma)^{0.31}(\Delta\rho/\rho_V)^{0.31}$$

and D = any characteristic length dimension of the equipment which is included solely to form dimensionless groups.

Example 4.9

A 50% sugar solution is to be boiled in an open pan. Calculate the boiling heat transfer coefficient and excess temperature to sustain a heat flux of 4.83×10^5 W/m^2 using McNelly's correlation. Liquid density = 1240 kg/m^3; viscosity at 100°C = 1.70 cP; specific heat = 3.4 kJ/kg K; surface tension at the boiling point = 0.0589 N/m; thermal conductivity = 0.555 W/m K; latent heat of vaporisation = 2.26×10^6 J/kg; vapour density at 100°C = 0.588 kg/m^3.

$(Pr)_L = (C_p\mu/k) = (3.4 \times 10^3 \times 1.7 \times 10^{-3}/0.555) = 10.41$

$(\bar{q}D/\lambda\mu) = (4.83 \times 10^5 D)/(2.26 \times 10^6 \times 1.7 \times 10^{-3}) = 125.7D$

$(PD/\sigma) = (1.013 \times 10^5 D/0.0589) = 1.72 \times 10^6 D$

$(\Delta\rho/\rho_V) = (1240 - 0.588)/0.588 = 2108$

$$(h_b D/k) = 0.225(10.41)^{0.69}(125.7D)^{0.69}(1.72 \times 10^6 D)^{0.31}(2108)^{0.31}$$
$$= 0.225(5.035)(28.09)(85.71)(10.73)D$$

$h_b = 29\ 266 \times 0.555 = 16\ 243\ W/m^2K.$

$\Delta T_x = 4.83 \times 10^5/16\ 243 = 30^oC.$

Example 4.10

For the same problem as in example 4.9, calculate the boiling heat transfer coefficient and excess temperature using Kutateladze's correlation.

$\psi = (\sigma/g\Delta\rho) = 0.0589/(9.81 \times 1240) = 4.84 \times 10^{-6}$

$a = 0.555/(3.4 \times 10^3 \times 1240) = 1.316 \times 10^{-7}$

$$(\bar{q}/a\lambda\rho_V) = 4.83 \times 10^5/(1.316 \times 10^{-7} \times 2.26 \times 10^6 \times 0.588)$$
$$= 2.762 \times 10^6$$

$(P/\sigma) = 1.013 \times 10^5/0.0589 = 1.72 \times 10^6$

$(Pr)_L = 10.41$ - from example 4.9

$$(h_b/k)\psi^{1/2} = 0.0007[(2.762 \times 10^6)(1.72 \times 10^6)(4.84 \times 10^{-6})]^{0.7} \times (10.41)^{-0.35}$$
$$= 0.0007(2.299 \times 10^7)^{0.7}(0.44)$$

$$h_b = 0.0007 \times 0.555 \times 1.423 \times 10^5 \times 0.44(4.84 \times 10^{-6})^{-0.5}$$
$$= 11\ 057\ W/m^2K.$$

$\Delta T_x = 4.83 \times 10^5/11\ 057 = 44^oC.$

Example 4.11

Based on the results obtained in examples 4.9 and 4.10 above, what will be the value of the Rohsenow constant for this problem ?

$(Pr)_L = 10.41$ - from example 4.9

$\psi = 4.84 \times 10^{-6}$ - from example 4.10

$\psi^{1/2} = 2.2 \times 10^{-3}$

$(\bar{q}/\lambda\mu) = 4.83 \times 10^5/(2.26 \times 10^6 \times 1.7 \times 10^{-3}) = 125.7$

$(C_p\Delta T_x/\lambda) = K(125.7 \times 2.2 \times 10^{-3})^{1/3}(10.41)^{1.7} = 34.96K$

$$\Delta T_x = 34.96K \times 2.26 \times 10^6/3.4 \times 10^3 = 23\ 239K$$

From example 4.9 - $\Delta T_x = 29.7°$, and $K = 0.00128$

From example 4.10 - $\Delta T_x = 43.7°$, and $K = 0.00188$.

(Experimental work on this system gave a value for the Rohsenow constant of 0.0015).

All of the above correlations are dependent on accurate values being available for the physical properties. The most unpredictable physical property in all the correlations is the surface tension. Anything which will affect this property (proteins, carbohydrates, surface active agents etc) will render the heat transfer correlations of dubious value for design purposes in a new system.

However, these correlations are of considerable use in being able to predict the effect of varying the process parameters on the pattern of heat transfer. For example, all the correlations show that

h_b is proportional to $(\bar{q})^a$, where a lies between 0.67 and 0.7

h_b is proportional to $(\sigma)^b$, where b lies between -0.2 and -0.5.

Prediction of Peak Heat Flux

The prediction of peak heat flux (alternatively 'burnout point') is important because of the rapid decrease in heat flux (and hence the heat transfer coefficient) once this point has been exceeded.

Rohsenow and Griffith [21] suggested the following relationship, which has been found to give an answer to ± 11%

$$\bar{q}_{max} = C\lambda\rho_V(\Delta\rho/\rho_V)^{0.6}(g_i/g)^{0.25}$$

g = acceleration due to gravity; g_i = value of an 'induced' gravitational field, e.g. in a centrifugal evaporator.

This expression is dimensional, and the constant C has the following values

(a) when all units are in the FPH system, C = 143

(b) when all units are in the SI system, C = 0.0121.

Kutateladze [22] suggests the following

$$\bar{q}_{max} = 0.16\lambda\rho_V(\sigma g\Delta\rho/\rho_V^2)^{0.25}$$

which is similar to the expression of Zuber [23]

$$\bar{q}_{max} = 0.131\lambda\rho_V(\sigma g\Delta\rho/\rho_V^2)^{0.25}(\Sigma\rho/\rho_L)^{0.5}$$

Example 4.12

Calculate the peak heat flux for boiling a 50% sugar solution in an open pan, using the correlations of (a) Rohsenow and Griffith, (b) Kutateladze, (c) Zuber. Asume that the physical properties are as stated in example 4.9.

(a) Rohsenow and Griffith.

$\rho_V = 0.588\ kg/m^3$; $\rho_L = 1240\ kg/m^3$; $\lambda = 2.26 \times 10^6\ J/kg$;

$(\Delta\rho/\rho_V) = 2108$ - from example 4.9

$$\bar{q}_{max} = 0.0121 \times 2.26 \times 10^6 \times 0.588(2108)^{0.6} \text{ and } g_i = g.$$

$$= 1.58 \times 10^6\ W/m^2.$$

(b) Kutateladze.

$$(\sigma g\Delta\rho/\rho_V{}^2) = (0.0589 \times 9.81 \times 1240/0.588^2) = 2072.3$$

$$\bar{q}_{max} = 0.16 \times 2.26 \times 10^6 \times 0.588(2072.3)^{0.25} = 1.43 \times 10^6\ W/m^2.$$

(c) Zuber.

$$(\Sigma\rho/\rho_L) = (1240 + 0.588)/1240 = 1.0005$$

$$\bar{q}_{max} = 0.131 \times 2.26 \times 10^6 \times 0.588(2072.3)^{0.25}(1.0005)^{0.5}$$

$$= 1.17 \times 10^6\ W/m^2.$$

From the above results, it can be seen that Rohsenow and Griffith agree to within the confidence limits of ± 11% with Kutateladze.

The major advantage of the Rohsenow and Griffith correlation is that it contains only physical properties which are not difficult to measure or estimate.

Boiling with Forced Convection

In order to apply the pool boiling correlations to combined boiling with forced convection, the total heat flux ($\bar{q}_t$) is separated into two components, one a boiling flux ($\bar{q}_b$) and the other a convection flux ($\bar{q}_c$).

The convection flux is calculated on the basis of normal liquid flow (assuming no boiling) using one of the accepted correlations to calculate the heat transfer coefficient (h_c), and

$$\bar{q}_c = h_c(t_s - t_1) = h_c\Delta T_x$$

Rohsenow [15] suggests that the value for $\bar{q}_t$ is the sum of the two fluxes for the region where boiling and forced convection are combined, e.g. in two-phase flow situations

$$\bar{q}_t = \bar{q}_b + \bar{q}_c.$$

Kutateladze [17] takes a slightly more pessimistic view

$$\bar{q}_t = \sqrt{(\bar{q}_b^2 + \bar{q}_c^2)}$$

Example 4.13

A 50% sugar solution is to be evaporated at atmospheric pressure in an evaporator having 10 x 30 mm diameter tubes. Each tube is 3.0 m long, and the flowrate of solution is 1645 kg/h. The solution enters the tubes at 72°C, and the wall temperature is 150°C.
(a) Calculate how far along the tubes boiling takes place, given that the heat transfer coefficient for convection is 500 W/m^2K and the boiling point of the solution is 102°C. (b) Calculate the total heat flux required to sustain a 10% evaporation rate and the mean heat transfer coefficient for the boiling section of the tubes. Liquid density = 1240 kg/m^3; mean liquid viscosity at 100°C = 1.7cP; specific heat of liquid = 3.4 kJ/kg K; thermal conductivity of the liquid = 0.555 W/m K; latent heat of vaporisation = 2.26×10^6 J/kg; vapour density = 0.588 kg/m^3.

(a) Heat required to raise the liquid from 72° to 102°C

$$q_c = [1645 \times 3400(102 - 72)]/3600 = 46\ 608 \text{ W.}$$
$$= h_c A \Delta t_{lm}$$
$$\Delta t_{lm} = [(150 - 72) - (150 - 102)]/\ln(78/48) = 61.8°$$

and $A = 46\ 608/(500 \times 61.8) = 1.508 m^2$

$= \pi D(10L)$ - for all of the 10 tubes.

thus $L = 1.6m$.

Boiling will take place 1.6 m along the tubes, leaving 1.4 m of each tube for the evaporation (combined boiling and convection) duty.

(b) Heat transfer by convection only $q_c = h_c A \Delta T_x$

A = area of tubes in which combined heat transfer takes place.

$$A = \pi \times 0.03 \times 1.4 \times 10 = 1.319 m^2 \text{ for all 10 tubes}$$
$$q_c = 500 \times 1.319(150 - 102) = 31\ 656 \text{ W.}$$

and $\bar{q}_c = 31\ 656/1.319 = 24\ 000\ W/m^2$.

Heat required to sustain 10% evaporation

$$q_b = (1645 \times 2.26 \times 10^6 \times 10\%)/3600 = 103\ 269 \text{ W}$$

and $\bar{q}_b = 103\ 269/1.319 = 78\ 293\ W/m^2$.

<u>Rohsenow.</u> $\bar{q}_t = 78\ 293 + 24\ 000 = 102\ 300\ W/m^2$

Mean heat transfer coefficient $h_t = 102\ 293/48 = 2130\ W/m^2K$.

<u>Kutateladze.</u> $\bar{q}_t = \sqrt{(78\ 293^2 + 24\ 000^2)} = 81\ 900\ W/m^2$

Mean heat transfer coefficient $h_t = 1710\ W/m^2K$.

For the flow of a vapour/liquid mixture through tubes, Davis and David [24] found that so long as the liquid wetted the tube wall, the overall heat transfer coefficient for combined boiling and forced convection (h_t) is empirically correlated by the equation

$$(h_tD/k) = 0.06(\rho_L/\rho_V)^{0.28}(Re.X)^{0.87}(Pr)_L^{0.4}$$

D = tube diameter; X = mass fraction of vapour; k = liquid thermal conductivity; $(Pr)_L$ = liquid Prandtl number; Re = liquid Reynolds number.

<u>Example 4.14</u>

For a 50% sugar solution operating under the same conditions as in example 4.13 above, calculate the combined heat transfer coefficient for the two-phase flow using the Davis and David correlation.

Based on a single tube

Liquid velocity $= (164.5 \times 4)/(3600 \times 1240 \times \pi \times 0.03^2)$

$= 0.052\ m/s$

Reynolds number Re $= (0.03 \times 0.052 \times 1240)/(1.7 \times 10^{-3})$

$= 1141$

$\rho_L/\rho_V = 1240/0.588 = 2108.8$

$(Pr)_L = (3400 \times 1.7 \times 10^{-3})/0.555 = 10.41$

$X = 0.1$

$(h_tD/k) = 0.06(2108.8)^{0.28}(1141 \times 0.1)^{0.87}(10.41)^{0.4}$

$= 80.47$

$h_t = 80.47 \times 0.555/0.03 = 1490\ W/m^2K$.

4.3 CLIMBING FILM EVAPORATOR

When a liquid is heated in bulk in order to evaporate water from it, the liquid is normally in contact with the heating medium for a long period of time. This can cause considerable damage to thermally labile materials, e.g. liquid foodstuffs such as milk and fruit juices. One well-established method of overcoming this problem is to cause the liquid to flow in a thin film over the heating surface. This thin film can be produced by applying a centrifugal force on the liquid to spread it over the heating surface (centrifugal evaporators),by spreading the liquid over the heating surface by a wiping action (wiped film evaporators) or by using the high velocity of the low density vapour to carry a thin film of liquid over the surface (climbing film evaporators). Heat transfer in such evaporators is complex: nucleate boiling can occur at the heating surface producing vapour which passes directly into the vapour stream; forced convection heat transfer through the film also occurs producing evaporation at the free liquid surface. The mechanisms are described by Lacey et al [25] and Coulson and Richardson [26]. This section considers the analysis of heat transfer in a climbing film evaporator.

In a climbing film evaporator a two-phase (liquid and vapour) mixture flows up a tube heated externally, normally by steam. As it rises evaporation produces a greater concentration of vapour in the mixture. Several workers [27-33] have shown that the tube wall to mixture heat transfer coefficient (h_m) depends on the vapour concentration. The general form of the relationship is $h_m/h_i = c(1/X_{tt})^n$, where h_i is the heat transfer coefficient to the single phase liquid under the same process conditions, c and n are constants. X_{tt} is the Lockhart-Martinelli parameter defined as $X_{tt} = (\rho_v/\rho_i)^{0.5}(\mu_i/\mu_v)^{0.1}\{(1-x)/x\}^{0.9}$ where ρ_v, μ_v, ρ_i, μ_i are the vapour and liquid densities and viscosities respectively, and x is the mass fraction of vapour in the mixture. Jamil and Lamb [34] summarised a number of experimental correlations which showed values of c ranging from 2.17 to 7.55 and values of n ranging from 0.328 to 0.75. For example Dengler and Addoms [27] found the relationship $h_m/h_i = 3.5[1/X_{tt}]^{0.5}$ for $0.25<1/X_{tt}<70$.

Since the heat transfer coefficient varies with vapour concentration it will not be constant over the whole tube length and the following procedure is suggested to allow for this variation in determining evaporator surface area for a particular duty using a single pass operation (i.e. no recirculation).

(a) Calculate an approximate flow rate per tube based on an assumed overall heat transfer coefficient (a figure of 2500W/m^{2}K is typical for this type of evaporator).

(b) Using this flow rate calculate h_i. If the Reynolds Number (Re) is greater than 10 000 the Dittus and Boelter correlation can be used (see Chapter 2.1). For laminar flow conditions (Re<2100) the relevant correlation from Chapter 2.1 can be used or, for 10^{-1} <Re.Pr.(D/L)< 10^4 Loncin and Merson [35] recommend: Nu = 3.65 + {0.0668Re.Pr.(D/L)/1 + 0.045 [Re.Pr.(D/L)]$^{2/3}$}, where Pr is the Prandtl Number, D and L are the tube diameter and length.

(c) Substituting h_i and values of the liquid and vapour properties into h_m and X_{tt} correlations derive the relationship between h_m and x.

(d) Taking values of x between zero (initial condition) and the required exit vapour concentration, calculate h_i amd hence the overall heat transfer coefficient (U) for each value of x. U is obtained as $1/U = 1/h_i + 1/h_s$, where h_s is the condensing steam heat transfer coefficient and wall resistance is negligible. In the absence of experimental data h_s can be taken as 8000W/m^{2}K. There are methods for predicting h_s (see, for example, Brown et al [36]) but, as has been reported [35,37],a number of factors can affect the accuracy of the correlations which tend to give a low value.

(e) Taking the average value for U for each interval change in x, calculate the tube length for each interval from a heat and mass balance. Summation of these lengths gives the total tube length required.

Example 4.15

Estimate the number and length of tubes required (maximum length 6m) in a climbing film evaporator carrying out the following duty. The liquid to be evaporated has a viscosity of 0.003kg/ms, density 1000kg/m^3, specific heat 4200J/kgK. The feed rate is 1000kg/hour and the feed is to be concentrated in a single pass from 12% (by mass) solids to 25%. The tube internal diameter is 30mm. The feed enters at boiling point (45°C) and the process steam is at 135°C.

Other physical properties are: $k_i = 0.9 W/m\,K$, $\rho_v = 0.04 kg/m^3$, $\mu_v = 1 \times 10^{-5} kg/ms$, latent heat ($\lambda$) = $2.3 \times 10^6 J/kg$.

(a) For an assumed tube length of 6m the surface area/tube = $6 \times \pi \times 0.03 = 0.5655 m^2$. For $U = 2500 W/m^2K$ the available heat exchange/tube = $2500 \times 0.5655 \times (135-45) = 1.272 \times 10^5 W$. This will allow evaporation of $1.27 \times 10^5/2.3 \times 10^6 kg/s$ = 199kg/hour. The feed/hour is 120kg solid + 880kg solute. The output from the evaporator/hour therefore has composition 120kg solid + 880kg (liquid + vapour). For 25% solids in the output liquid there must be 480kg solution and hence 520kg vapour/hour. Therefore 520/199 tubes are required: so estimate on 3 tubes, each with a feed of 333kg/hour.

(b) To calculate h_i, first determine the liquid velocity for single phase flow = volumetric flow rate/X-sectional area = $\{333/(3600 \times 1000)\}/\{\pi \times 0.03^2/4\}$ = 0.1309m/s. Hence Re = $0.1309 \times 0.03 \times 1000/0.003$ = 1309. Since Re < 2100 the flow is laminar. Pr = $4200 \times 0.003/0.9$ = 14. (Re.Pr.D/L) = $1309 \times 14 \times 0.03/6$ = 91.6. Therefore $Nu = 3.65 + (0.0668\, Re.Pr.D/L)/1 + 0.045(Re.Pr.D/L)^{2/3} = 6.85$. $h_i = Nu(k_i/D) = 6.85 \times 0.9/0.03 = 206 W/m^2K$.

(c) $X_{tt} = (0.04/1000)^{0.5}(0.003/1 \times 10^{-5})^{0.1}\{(1-x)/x\}^{0.9} = 0.0112\{(1-x)/x\}^{0.9}$. Using $h_m = 3.5h_i\,(1/X_{tt})^{0.5}$ gives $h_m = 3.5 \times 206 \times 0.0112^{-0.5}\{x/(1-x)\}^{0.45} = 6813\,\{x/(1-x)\}^{0.45}$

(d) Values of h_m from this equation are tabulated below. The values for U are calculated using $h_s = 8000 W/m^2K$ and hence $1/U = (1/h_m) + 0.000125$. The maximum value of x is exit vapour rate/total rate = 520/1000 = 0.52.

(e) U_{av} is calculated as the average U for each interval in x. ΔL, the tube length for each Δx, is obtained by equating the heat exchange in that section to the mass evaporated x latent heat. This gives $U_{av}(\pi \times 0.03 \times \Delta L)(135-45) = (333/3600)(\Delta x)(2.3 \times 10^6)$. This equation simplifies to $\Delta L = 25082(1/U_{av})\Delta x$.

x	$(x/1-x)$	h_m	$1/U(\times 10^4)$	$1/U_{av}(\times 10^4)$	ΔL
0	0	206	49.79		
				41.97	0.105
0.001	0.001	304	34.14		
				25.63	0.257
0.005	0.00503	630	17.12		
				14.99	0.188
0.01	0.0101	862	12.85		
				11.28	0.283
0.02	0.0204	1182	9.710		
				8.241	0.620
0.05	0.0526	1811	6.772		
				5.984	0.750
0.1	0.1111	2535	5.195		
				4.592	1.152
0.2	0.25	3651	3.989		
				3.694	0.927
0.3	0.4286	4653	3.399		
				3.205	0.804
0.4	0.6667	5677	3.011		
				2.839	0.854
0.52	1.0833	7063	2.666		
				$\Sigma\Delta L$ =	5.94

Thus the required tube length is 5.94m, and the feed should be split between three such tubes. If the tube length had been greater than the stipulated maximum of 6m the calculation would be repeated with a larger number of tubes.

Example 4.16

The feed of example 4.15 is to be concentrated to 52% solids in the same evaporator. Establish a suitable process for this and the time to produce 100kg of concentrated product.

Three possible procedures will be considered: (a) the required quantity of feed could be passed through the evaporator, the concentrate batched and then returned to the evaporator until the desired concentration is reached, or (b) concentrate could be continuously recirculated, with fresh feed added to keep a constant feed rate to the evaporator (1000kg/hour as previously) or (c) the feed rate could be reduced to allow a longer residence time in the evaporator and hence achieve the desired concentration in one pass.

(a) After one pass the concentrate is 25% solids (see example 4.15). If this is returned to the evaporator at 1000kg/hour, solids into and out of the evaporator = 250kg/hr. Since the evaporation rate is 520kg/hour the solids concentration in the output liquid after two passes will be (250/480) x 100 = 52%. Therefore two passes will be required. The total amount of water to be evaporated in the two passes is (100 x 0.52)(0.88/0.12 - 0.48/0.52) = 333.3kg. The evaporation time is therefore (333.3/520) x 60 = 38.5 mins. Since the liquid to be batched after the first pass has a mass of (100 x 0.52) x (100/25) = 208kg a concentrate collection vessel of this capacity is required for this operation.

(b) With continuous recirculation the same flow rate into the evaporator is maintained and hence the evaporation rate will be the same as in (a). Therefore the evaporation time will be the same (38.5mins). A concentrate receiver of 100kg ($0.1m^3$) capacity will be required with arrangements for recirculation from this receiver.

(c) Reducing the feed rate reduces the heat transfer coefficient and the procedure used in example 4.15 can be used to calculate the feed rate to give a concentration of 52% in the exit liquid from a 6m tube. (If the feed rate to the evaporator is F, the solids leaving and entering = 0.12F, thus the liquid rate leaving = 0.12F/0.52 = 0.231F. The vapour rate leaving = F - 0.231F and hence x = F-0.231F/F = 0.769, i.e. it is independent of F). This calculation gives a feed rate per tube of approximately 210kg/hour, or 630 kg/hour for the 3 tubes. The evaporation rate (see above) = 0.769 x 630 = 484kg/hour (cf 520kg/hour at the faster feed rate) and the evaporation time is (333.3/484) x 60 = 41.3mins. Although this system of operation is simpler than (a) or (b) it would be necessary to ensure that satisfactory flow of the film was obtained at the lower velocities used.

The above calculations make no allowance for increasing viscosity as concentration progresses. Where it is known that this is significant, higher values of μ_i can be used in computing h_i as x increases.

4.4 AIR CONDITIONING

Air conditioning refers to the control of the temperature and humidity of the air within an operating area or storage area. These conditions may be required from either a processing or storage point of view, or may be required for the comfort of the workforce.

Temperature. The temperature of the air within a working area is affected by the following factors

(1) the type and number of occupants and their workload

(2) the type and size of heat producing equipment, e.g. motors, lights, stock movements etc

(3) the construction of the walls, roof, windows, doors etc

(4) outside weather conditions, e.g. heat gains due to outside temperature, solar radiation, losses in cold weather etc.

Humidity. The humidity of the air in a working area is affected by moisture gains, or losses, due to

(1) process operations

(2) occupants

(3) changes of air from outside and from other areas (due to opening doors etc).

In order to be able to specify plant for air conditioning, the heat and moisture gains to the area under consideration must be calculated, and the plant specified for the worst conditions expected, e.g. the maximum heat gain can be expected in mid-Summer. In addition, the size of plant required will be affected by the temperature and humidity limits which can be tolerated. Thus, if a wide latitude of temperature can be tolerated, simple ventilation of the area may be sufficient.

The most common method used for air conditioning is by circulation and treatment of a portion of the air drawn from the area. This portion of air may be treated to reduce both temperature and humidity before redistribution back into the area. It is also common practice to reject part of the circulating air and replace this with filtered fresh air to repress the build-up of, for example, bacteria, body odours etc.

The actual scheme used for any particular problem can consist of strict control of the environment, or may simply involve replacement of some of the air in the area with fresh air from outside (ventilation).

Peak Heat Gains

Calculation of the maximum heat gain to be removed by a treatment plant involves the summation of all possible sources of heat, and data is widely published on the following:

(a) thermal transmittance of building materials ('U' values in W/m^2K)

(b) solar heat gains for different latitudes, time of day and time of year

(c) heat generated by occupants, motors, lights etc.

A selection of values for the above are given in tables 4.1 to 4.5 inclusive.

Table 4.1 [38]

Maximum Solar Heat Penetration through Walls in Clear Weather (W/m^2)

Inside temperature = Outside temperature

Latitude = 16° N or S

	Thickness (mm)	E and W Facing	SE and SW* Facing	South* Facing
Brick, solid unplastered	105	142	74	13.6
	220	94	53	9.4
Brick, solid 16 mm plaster	105	117	62	11.7
	220	88	47	8.8
Brick, cavity 16 mm plaster	260	61	31	6.7
	375	48	30	5.4
Concrete	150	127	75	12.7
	200	110	58	10.4
Sheet asbestos, flat	3.5	319	165	30
corrugated	3.5	414	214	39
	5.0	296	144	26
Sheet glass, bare windows		1050	625	221

*Note: Reverse bearings for Southern latitudes.

Table 4.2 [38]

Maximum Solar Heat Penetration through Roofs in Clear Weather (W/m^2)

Flat: asphalt on 150 mm concrete	167
as above with 16 mm plaster	108
Pitched: corrugated asbestos (5 mm)	420
tiles with boards and felt	102

Table 4.3 [38]

Solar Intensity Correction Factors for Latitude.

Latitude (N or S)	Horizontal Surface	E & W Wall	SE & SW* Wall	South* Wall
16	1.0	1.0	1.0	1.0
24	0.98	1.0	1.13	1.57
32	0.94	1.0	1.27	2.41
40	0.88	0.99	1.37	3.24
48	0.79	0.96	1.45	3.91
56	0.68	0.93	1.50	4.41
64	0.56	0.88	1.52	4.72

*Note: Reverse Bearing for Southern Latitudes.

Table 4.4 [39]

Thermal Transmittance of Building Materials (U-values) (W/m^2K)

	Thickness (mm)	U-value
Walls		
Brick, solid, unplastered	105	3.3
	220	2.3
solid, 16mm plaster	105	3.0
	220	2.1
cavity, 16mm plaster	260	1.5
	375	1.2
Concrete, cast	150	3.5
	200	3.1
Single Windows		5.6
Double Windows, 6mm airspace		3.4
Asbestos, flat	3.5	5.3
corrugated	3.5	6.9
	5.0	4.8
Roofs		
Flat, asphalt on 150mm concrete		3.4
as above with 16mm plaster		2.2
Pitched, corrugated asbestos	5.0	4.3
tiles on boards & felt		1.3

Table 4.5 [39]

Heat given Off from People (W)

Seated at rest	115
Engaged on Light Work	140
Engaged on Light Bench Work	235
Engaged on Medium Work (say walking at 3 mph)	265
Heavy Work (4mph)	440

The values given in tables 4.1 and 4.2 are based on the average time of intensity and thermal resistance of the building materials. The values given are peak heat flows which can be expected to penetrate during August in northern latitudes, and it should be noted that the peak due to solar radiation does not necessarily occur at the time of maximum shade temperature since the effect of solar radiation is that of a slow build-up. In temperate zones (40° - 60° N or S), the peak due to solar radiation occurs at approximately 1600 hours local; peak shade temperature is usually earlier, at about 1330 - 1430 hours local.

Simple Ventilation Requirements

Simple ventilation for heat removal involves replacing some of the air in an area with air from outside. Obviously, the temperature inside the area can never become lower than the outside temperature, and it is uneconomic to try to cool an area to within less than 3°C of the outside temperature by ventilation only.

Even to cool within 3°C will mean a large circulation rate of outside air, and it is more normal to use an 'approach' temperature of 5° to 7°C.

Example 4.17

A building, 40 m x 20 m x 7 m high is occupied by 20 people working at benches. The peak heat gain in mid-Summer is calculated to be 6.0 kW. If the outside shade temperature is 20°C, what will be the ventilation requirements for maintaining a temperature of 25°C inside the building ? Mean specific heat of the air = 1.0 kJ/kg K; mean specific volume of the air = 0.795 m^3/kg.

Total heat gain = 6000 W; temperature difference = 5°

$$q = mC_p\Delta t$$

q = heat gain; m = mass flowrate of air; Δt = temperature difference.

$$q = 6000 = m(1000 \times 5)$$

and $m = 1.2$ kg/s.

The ventilation rate = 1.2 x 0.795 = 0.954 m^3/s.

(The ventilation rate with an 'approach' temperature of 3° would be 66.7% greater).

Example 4.18

Calculate the maximum solar heat gain through a 5 mm thick corrugated asbestos sheeted wall, 30 m long x 6 m high, with 5 windows, 3 m x 1.25 m, facing East in a latitude of 32° N.

Area of asbestos = (30 x 6) - (5 x 3 x 1.25) = 161.25 m^2

Area of windows = 18.75 m^2

From table 4.1, solar gains at 16° latitude

for corrugated asbestos 5 mm thick = 296 W/m^2

for bare window glass = 1050 W/m^2

From table 4.3, correction factor for 32° N = 1.0.

Heat gains:

Asbestos wall = 161.25 x 296 = 47 730 W

Windows = 18.75 x 1050 = 19 687.5 W

Total maximum gain = 67 417.5 W = 67.4 kW.

Air Conditioning requirements

Where full air conditioning is required, i.e. control of both temperature and humidity, then the load on the treatment plant is determined as follows:

Temperature. The peak heat load to be removed is calculated from all sources

(a) due to temperature differences through the walls, roof etc.

(b) due to solar radiation at peak rates

(c) due to equipment, stock, occupants etc.

As a guide to the peak heat gain in mid-Summer, this method of calculation will give an over-estimate, and more precise calculations taking account of the slow build-up of solar heat can be made using methods described in the literature [38, 39]. Maximum Summer shade temperatures based on worldwide meteorological data can be found from the same sources.

Once the peak heat load has been calculated, the air circulation rate can be obtained once the air temperature difference which can be tolerated has been determined.

For areas occupied by people, 5° difference between circulated air and controlled temperature is normal, lower temperatures can be uncomfortable causing cold draughts. A larger difference can be tolerated in storerooms, thus reducing the air circulation rate.

Humidity. The peak moisture gains in the area are calculated

(a) due to the occupants, (see table 4.6)

(b) from air changes due to opening doors etc.

(c) from process operations.

A moisture balance can then be performed to determine the amount of water to be removed from the circulating air (the air rate determined from the temperature requirements above).

Table 4.6 [40]

Moisture Production by Human Occupants

at Rest	-- 0.091 kg/hour
working hard	-- 0.272 kg/hour
average	-- 0.182 kg/hour

When air is cooled below its dew point, moisture condenses out, and air conditioning plants typically use finned tubes with either brine or direct expansion of refrigerant inside the tubes. Typical air temperatues leaving such finned tubes are of the order of 0° to +5°C, which is normally well below the dew point. This type of refrigerated system not only dehumidifies the air but reduces the temperature.

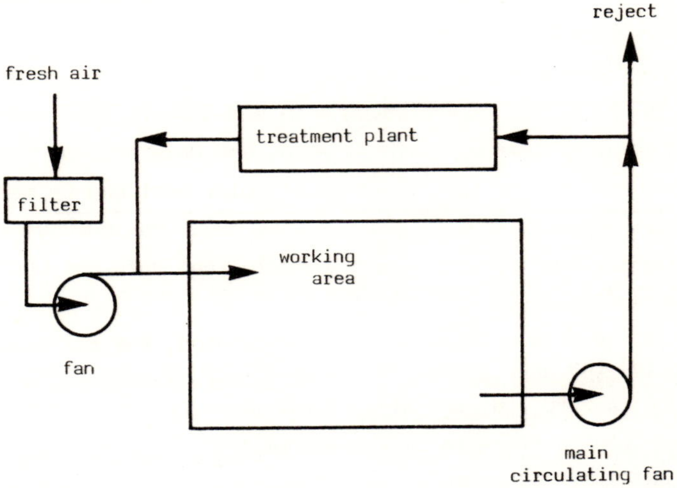

Figure 4.6 - A typical air conditioning scheme.

Because the temperature of the air is reduced to about 5°C, reheating of the air is necessary, even in Summer, and rehumidification is quite often needed, depending on the standard of the control system installed.

Example 4.19

The plan view of a room (B) and the surrounding areas are shown in figure 4.7. The whole area is situated at a latitude of 56° N, and the temperatures shown in figure 4.7 are controlled.

Room B has a ground area of 30 m x 15 m, walls 6 m high, and the construction is as follows:

Outside wall - 260 mm cavity brick with 16 mm plaster, facing SE.

Inside walls - 105 mm solid brick with 16 mm plaster.

Roof - flat, 150 mm concrete with asphalt.

Floor - 200 mm concrete.

Air changes through door = ½ change/h; occupants = 3 (on medium work); lights = 20 x 80 W; no windows. Assume that the mean ground temperature = 15°C.

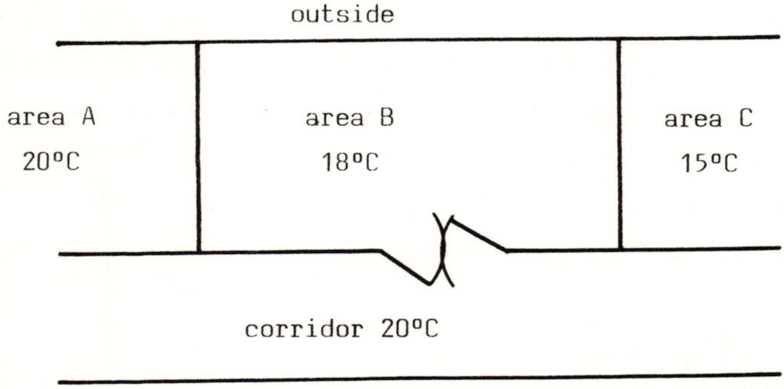

Figure 4.7 - Plan view of area in example 4.19

Calculate (a) the peak Summer heat load if the maximum outside shade temperature is 25°C, (b) the size of circulating fan required for a 5° temperature difference in the circulated air. Mean specific heat of the air = 1.0 kJ/kg K; mean specific volume = 0.8 m^3/kg.

(a) Peak Summer heat gains.

(1) Solar heat gains

Source	Area (m^2)	Gain (W/m^2)*1	Correction*2	Gain (W)
Outside wall	180	31	1.50	8370
Roof	450	167	0.68	51102

Note: *1 - table 4.1, *2 - table 4.3

(2) Conduction gains

Source	Area (m^2)	U value*3	Δt°	Gain (W)
Outside wall	180	1.5	7	1890
Roof	450	3.4	7	10710
from Room A	90	3.0	2	540
from Room C	90	3.0	-3	-810 (loss)
Corridor (+ door)	180	3.0	2	1080
Floor	450	3.1	-3	-4185 (loss)

Note: *3 - table 4.4

(3) Air changes

Volume of room = (30 x 15 x 6) = 2700 m^3

Changes through door = 2700/2 = 1350 m^3/h

Weight rate of change = 1350/(0.8 x 3600) = 0.469 kg/s

Gain due to air changes = (0.469 x 1000 x 2) = 938 W

(4) Lights

Gain = 20 x 80 = 1600 W.

(5) Occupants - 3 men engaged on medium work. From table 4.5 heat given off from each person = 256 W.

Gain = 3 x 265 = 795 W.

Total Summer heat gain from all sources = 72030 W, i.e. 72.0 k

(b) Size of circulating fan.

Temperature difference of circulating air = 5°

Heat gain $q = mC_p\Delta t$

thus 72030 = m(1000 x 5), and m = 14.41 kg/s

Volumetric rate = 14.41 x 0.8 = 11.5 m^3/s.

Example 4.20

In example 4.19 above, what will be the net Summer moisture gains to Room B given the following information:

Room B is to be controlled at 18°C, 50% humidity; outside air is at 25°C, 78% humidity; corridor is at 20°C, 80% humidity; fresh air makeup is 10% of the circulation rate.

The sources for the moisture gains are

(1) Occupants

From table 4.6, average respiration rate = 0.182 kg/h

Moisture gain = 3 x 0.182 = 0.546 kg/h.

(2) Air changes from the corridor

Humidity of the air in the corridor = 0.0118 kg/kg air - from the psychrometric chart (figure 4.2, section 4.1).

Humidity of the air in Room B (18°C, 50% humidity) = 0.0065 kg/kg.

Air change rate = 0.469 kg/s - from example 4.19

Gain = (0.0118 - 0.0065)(0.469 x 3600) = 8.949 kg/h.

(3) Fresh air makeup

Humidity of fresh air = 0.0136 kg/kg air (25°C, 78% humidity)

Circulation rate = 14.41 kg/s - from example 4.19

Gain = (0.0136 - 0.0065)(14.41 x 10% x 3600) = 36.83 kg/h.

Total net Summer moisture gains from all sources

= 0.546 + 8.949 + 36.83 = 46.325 kg/h, i.e. 0.013 kg/s.

<u>Example 4.21</u>

The circulating air in examples 4.19 and 4.20 above, is to be treated after extraction from the room by passing it over refrigerated,

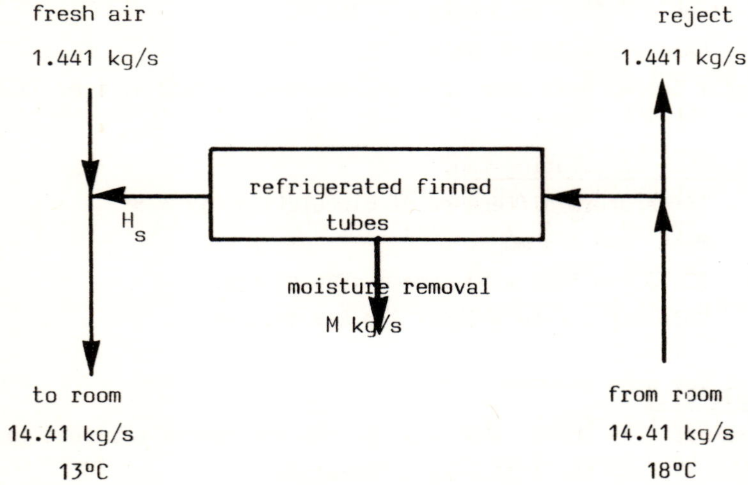

Figure 4.8 - Schematic diagram of system in example 4.21

finned tubes, the system being shown schematically in figure 4.8.

(a) What temperature must the air leave the refrigerated cooler in order to balance the moisture gains found in example 4.20 ? (b) what will be the heat load on the cooler ? (c) what will be the re-heat load on the air before returning it to the room ?

(a) An overall balance for moisture will be

(room gains) = (10% rejection) + (removal on cooler)

thus 0.01287 = (1.441 x 0.0065) + M

Removal rate on the cooler M = 0.0035 kg/s

The air must be cooled below the dew point for condensation (and hence removal), thus, the air leaving the cooler will be saturated.

If H_s = saturation humidity of the air leaving the cooler

$M = (14.41 \times 90\%)(0.0065 - H_s) = 0.0035$ kg/s

and $H_s = 0.00623$ kg/kg air.

From the psychrometric chart, this corresponds to a temperature of 7.3°C.

(b) Heat load

The total heat load on the cooler will be made up of the sensible heat in cooling the air from 18°C to 7.3°C and the latent heat of condensation of the water removed.

$C_p = 1.0$ kJ/kg K; $\lambda = 2.3 \times 10^6$ J/kg.

Total heat load = $(12.969 \times 1000 \times 10.7) + (0.0035 \times 2.3 \times 10^6)$

= 146 818.3 W, i.e. 147 kW.

(c) Re-heat load

Air off cooler = 7.3°C; fresh air = 25°C; air is to be admitted back to Room B at 13°C (Δt = 5°).

A heat balance above a datum of 0°C will give

$(12.969 \times 7.3C_p) + (1.441 \times 25C_p) + (\text{re-heat}) = (14.41 \times 13C_p)$

Re-heat load = 57 kW.

Note that the re-heater would be sized for Winter conditions.

It has been assumed in example 4.21 that all the circulating air (less the reject portion) would be passed over the cooler to reduce both temperature and humidity, and that the temperature of the air would have been controlled to give a moisture balance over the whole system.

A better technique, which can show considerable energy savings, is to treat only a portion of the circulating air using the cooler, the rest of the air being diverted through a by-pass for mixing with the air from the cooler and the fresh air makeup. The amount of air diverted through the by-pass is usually controlled manually using simple shutter valves.

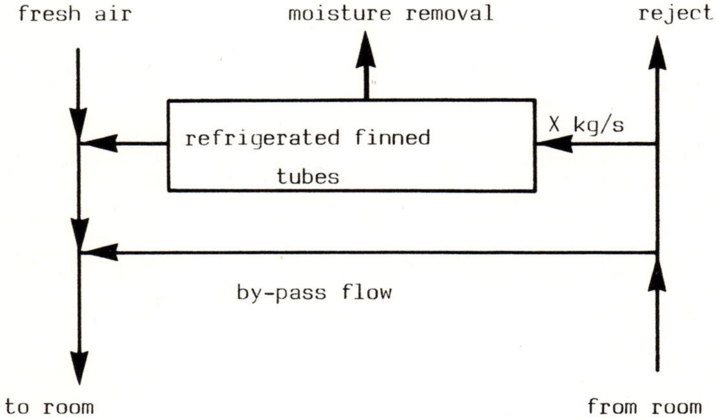

Figure 4.9 - Air conditioning system with by-pass

Example 4.22

The circulating air in example 4.19 and 4.20, is to be treated in a refrigerated cooler using a by-pass system. The air passing through the cooler leaves at +5°C.

(a) what will be the by-pass flowrate required to balance the moisture gains ? (b) what will be the by-pass rate required to balance the heat gains ?

The total mass flowrate of air over the cooler and through the by-pass = 12.969 kg/s - from example 4.21.

Let the airflow over the cooler = X kg/s

The overall moisture balance will be the same as in example 4.21

$M = 0.0035$ kg/s

Humidity of air extracted from the room = 0.0065 kg/kg air

Moisture into cooler = 0.0065X kg/s

Saturation humidity of air at 5°C = 0.0054 kg/kg air.

thus $0.0065X = 0.0054X + 0.0035$

and $X = 0.0035/(0.0065 - 0.0054) = 3.182$ kg/s

By-pass air flowrate = (12.969 - 3.182) = 9.8 kg/s.

(b) Peak heat gain = 72.03 kW - from example 4.19

Let the airflow over the cooler = Y kg/s

Temperature difference Δt = (18 - 5) = 13°

$q = YC_p \; \Delta t = Y \times 1000 \times 13$

and $Y = 72\,030/(1000 \times 13) = 5.54$ kg/s

By-pass air flowrate = (12.969 - 5.54) = 7.4 kg/s.

The conclusions to be drawn from examples 4.21 and 4.22 are as follows

(a) The heat removal load on the cooler is less using a by-pass system, 72.0 kW compared to 146.8 kW treating all the circulating air.

(b) With this particular system, the flowrate over the cooler is determined by the heat balance and re-humidification of the air would be required to balance the moisture.

(c) A simultaneous balance of heat removal and moisture removal shows that a flowrate of 6.15 kg/s over the cooler with an exit temperature of 6.35°C would exactly satisfy both balances.

(d) The re-heat load with an airflow rate of 5.54 kg/s over the cooler would be 10.12 kW compared to 56.63 kW in example 4.21.

In practice, the Summer gains would be used to determine the heat load (cooling) required and a by-pass used which could be controlled to balance heating and cooling conditions throughout the year.

REFERENCES

1. R.B. Keey, Drying, Principles and Practice, (Pergamon, Oxford, 1972)
2. W.B. Van Arsdel et al, Food Dehydration, Vols 1 and 2, 2nd Ed., (AVI, Westport, 1973)
3. Adapted by R.B. Keey from J.M. Coulson and J.F. Richardson, Chemical Engineering, Vol.1, 2nd Ed., (Pergamon, Oxford, 1970)
4. W.L. McCabe and J.C. Smith, Unit Operations of Chemical Engineering, 3rd Ed., (McGraw-Hill, New York, 1976)

5. W.R. Marshall and S.J. Friedman, 'Drying', in Chemical Engineers' Handbook (ed. J.H. Perry), 3rd Ed., (McGraw-Hill, New York, 1950)
6. G.D. Saravocos and S.E. Charm, 'Effect of surface-active agents on dehydration of fruits and vegetables', Fd Technol., Champaign, 16 (1962) 91
7. A.J. Scott, in Fundamental Aspects of the Dehydration of Foodstuffs, (Soc. Chem. Ind., London, 1958)
8. H.G. Kessler, Technische Hochschule Dissertation (1961), cited by R.B. Keey (see 1 above)
9. S.E. Charm, Fundamentals of Food Engineering, 2nd Ed., (AVI, Westport, 1971)
10. O. Krischer, Die wissenschaftlichen Grundlagen der Trocknungstechnik, 2nd Ed., (Springer-Verlag, Berlin, 1963)
11. F.W. Bakker-Arkema, 'Theoretical aspects of the drying of forage wafers', Diss. Abs., 26 (1965) 241
12. A.C. Jason, 'A study of evaporation and diffusion processes in the drying of fish muscle' in Fundamental Aspects of the Dehydration of Foodstuffs, (Soc. Chem. Ind., London, 1958)
13. H.A.C. Thijssen, 'Effect of process conditions in freeze drying on retention of volatile components' in Advances in Preconcentration and Dehydration of Foods, (ed. A. Spicer, Appl. Sci. Publ., London, 1974)
14. M. Karel et al, Principles of Food Science, PartII, Physical Principles of Food Preservation, (Marcel Dekker, New York, 1975)
15. W.M. Rohsenow, 'A method of correlating heat-transfer data for surface boiling liquids', Trans.Am.Soc.mech.Engrs, 74 (1952) 969
16. R.I. Vachon, G.H. Nix and G.E. Tanger, 'Evaluation of constants for the Rohsenow pool-boiling correlation', Jl Heat Transfer, 90 (1968) 239
17. S.S. Kutateladze, Heat Transfer in Condensation and Boiling, 2nd Ed., (Leningrad, 1952)
18. C.H. Gilmour, 'Performance of vaporizers: Heat transfer analysis of plant data', Chem.Engng Prog.Symp.Ser., 55(29) (1959) 67
19. J.W. Palen and J.J. Taborek, 'Refinery kettle reboilers: Proposed method for design and optimisation', Chem.Engng Prog., 58(7) (1962) 37

20. M.J. McNelly, 'A correlation for the rates of heat transfer to nucleate boiling liquids', J.imp.Coll.chem.Engng Soc.,7 (1953) 18
21. W.M. Rohsenow and P. Griffith, 'Correlation of maximum heat flux data for boiling of saturated liquids', Chem.Engng Prog.Symp. Ser., 52(18) (1956) 47
22. S.S. Kutateladze, 'Boiling heat transfer', Int.J.Heat Mass Transfer, 4 (1961) 31
23. N. Zuber, 'On the stability of boiling heat transfer', Trans. Am.Soc.mech.Engrs, 80 (1958) 711
24. E.J. Davis and M.M. David, 'Two-phase gas-liquid convection heat transfer', Ind.Engng Chem.Fundamentals, 3 (1964) 111
25. P.M.C. Lacey et al, Symposium on Two Phase Flow, Paper 1, I.Mech.E., London (1962)
26. J.M. Coulson and J.F. Richardson, Chemical Engineering, Vol.2, 2nd Ed., (Pergamon, Oxford, 1968)
27. C.E. Dengler and J.N. Addoms, 'Heat transfer mechanism for vaporization of water in a vertical tube', Chem.Engng Prog.Symp. Ser., 52 (1956) 95
28. S.A. Guerrieri and R.D. Talty, 'Heat transfer to organic liquids in single-tube, natural circulation, vertical-tube boilers', Chem.Engng Prog.Symp.Ser., 52 (1956) 69
29. V.E. Schrock and L.M. Grossman, 'Local pressure gradients in forced convection vaporization', Nucl.Sci.Engng,12 (1962) 474
30. R.M. Wright et al, 'Downflow boiling of water and n-butanol in uniformly heated tubes', Chem.Engng Prog.Symp.Ser., 61 (1965) 220
31. L. Pujol and A.H. Stenning, in 'Co-current Gas-Liquid Flow', Symp.Ser.1, Can.Soc.Ch.Eng. (1969)
32. J.G. Collier et al, 'Heat transfer to two-phase gas-liquid systems, Part II', Trans.Inst.chem.Engrs, 42 (1964) T127
33. J.B. Chaddock and H. Brunemann, Report HL-113, Duke University, North Carolina (1967)
34. M.Jamil and J. Lamb, 'Heat transfer in low rate evaporation from thin films', Proc.Int.Symp.Heat and Mass Transfer in Food Engng, Wageningen (1972) D7.1
35. M. Loncin and R.L. Merson, Food Engineering, (Academic Press, New York, 1979)

36. G.G. Brown *et al*, *Unit Operations*, (Chapman and Hall, London, 1950)
37. J.H. Perry (Ed.), *Chemical Engineers' Handbook*, 3rd Ed., (McGraw-Hill, New York, 1950)
38. ASHRAE Handbook, *Fundamentals*, (Am.Soc.Heatg Refrig.Air Conditioning Engrs., New York, 1977)
39. IHVE Guide, *Books A, B and C (S.I. Units)*, (Inst.Heatg Ventilg Engrs, London,1970)
40. ASHRAE Handbook, *Equipment*, (Am.Soc.Heatg Refrig.Air Conditioning Engrs, New York, 1975)

5 SEPARATION METHODS

5.1 FILTRATION

For the flow of a fluid through a porous bed of particles, the relationship proposed by Darcy [1] and confirmed by a number of workers, including Carman [2] can be written as

$$(dV/d\theta) = (A\Delta P)/(\alpha\mu L) = \text{rate of flow of fluid}$$

V = volume of fluid; θ = time; A = cross-sectional area of the bed normal to the fluid flow; ΔP = pressure drop across the bed; L = thickness of the bed; μ = viscosity of fluid; α = specific resistance of the bed = $(\text{permeability})^{-1}$.

This relationship can be applied to the filtration of a solid suspension, but since a filter medium is usually required to support the cake, this extra resistance to fluid flow must be taken into account. If the resistance of the filter medium and initial layers of cake are taken to be equivalent to a thickness of cake L_0, then assuming that the filter cake builds up at a rate proportional to the flow of filtrate

$$(dV/d\theta) = (A\Delta P)/[\alpha\mu(L + L_0)]$$

and V now applies to the flow of separated fluid, i.e. filtrate.

Incompressible Filter Cakes

A filter cake is assumed to be incompressible when the voidage, specific surface and specific resistance remain constant for all conditions of pressure drop and throughout the thickness of cake. For an incompressible cake

$$vV = LA = \text{volume of the filter cake for filtrate volume } V$$

v = volume of cake per unit volume of filtrate, and the basic expression for the instantaneous rate of filtration at any point in the process will be

$$(d\theta/dV) = (\alpha\mu vV)/(A^2\Delta P) + (\alpha\mu L_0)/(A\Delta P)$$

Constant rate filtration. For a constant rate of filtration $(dV/d\theta) = V/\theta$, and the basic expression for $(d\theta/dV)$ becomes

$$\theta = [(\alpha\mu v)/A^2\Delta P)]\, V^2 + [(\alpha\mu L_0)/(A\Delta P)]\, V$$

θ = time of filtration to produce a volume of filtrate V.

Constant pressure filtration. Integration of the basic expression for $(d\theta/dV)$ gives

$$\theta = [(\alpha\mu v)/(2A^2\Delta P)]V^2 + [(\alpha\mu L_o)/(A\Delta P)]V$$

For both modes of operation (constant rate and constant pressure) a plot of (θ/V) against (V) will be a straight line, and will allow experimental determination of the specific resistance of the filter cake, and also evaluation of the equivalent cake thickness of the filter medium and initial layers of cake (L_o). However, although reliable results for the value of α can be obtained in the laboratory, results for L_o are less reliable since this value depends on the startup conditions of the filter operation. Values obtained for L_o in the laboratory can only give an indication of the order of magnitude which can be expected on a full-scale plant.

Example 5.1

Filtration of a suspension in the laboratory using a pressure of 2.0 bar gave the following results

Time (min)	Volume of filtrate collected (m^3)
5.0	1.0×10^{-3}
30.0	4.2×10^{-3}

The area of filtration was 35.0 cm^2. Determine (a) the specific resistance of the filter cake (b) the equivalent cake thickness of the filter cloth (c) how long it would have taken to collect 5.0 l of filtrate. Viscosity of filtrate = 1.0 cP; value of v = 0.2.

(a) $(\theta/V) = [(\alpha\mu v)/(2A^2\Delta P)]V + [(\alpha\mu L_o)/(A\Delta P)]$

or $(\theta/V) = aV + b$

for the first reading

$$(5 \times 60)/(0.001) = 0.001a + b$$

for the second reading

$$(30 \times 60)/(0.0042) = 0.0042a + b$$

subtracting one from the other

$$(4.286 \times 10^5 - 3.0 \times 10^5) = 0.0032a$$

thus $a = 4.018 \times 10^7 = (\alpha\mu v)/(2A^2\Delta P)$

and $\alpha = (a2A^2\Delta P)/(\mu v)$

$= (4.018 \times 10^7 \times 2 \times [35 \times 10^{-4}]^2 \times 2 \times 10^5)/(10^{-3} \times 0.2)$

$= 9.84 \times 10^{11}$

(b) Using the experimental results and substituting for a in the general expression

$$(5 \times 60)/\ 0.001 = (4.018 \times 10^7 \times 0.001) + b$$

$$b = 2.6 \times 10^5 = (\alpha\mu L_o)/(A\Delta P)$$

and

$$L_o = (bA\Delta P)/(\alpha\mu)$$

$$= (2.6 \times 10^5 \times 35 \times 10^{-4} \times 2 \times 10^5)/(9.84 \times 10^{11} \times 10^{-3})$$

$$= 0.185 \text{ m}$$

(c) The relationship for the time of filtration has now been established as

$$\theta = (4.018 \times 10^7)V^2 + (2.6 \times 10^5)V$$

thus, the time to collect 5.0 l will be given by

$$\theta = (4.018 \times 10^7 \times 0.005^2) + (2.6 \times 10^5 \times 0.005)$$

$$= 1004.5 + 1300.0$$

$$= 2304.5 \text{ s.}$$

Thus, (a) $\alpha = 9.84 \times 10^{11}$ (b) $L_o = 0.185$ m (c) time to collect 5.0 l = 2304.5 s = 38.4 min.

Washing of the filter cake. Two types of washing processes are used in filtration

(a) Simple washing, where the washing fluid is fed through the same ports or channels as the suspension. In this case

Rate of washing = Final rate of filtration

$$R_w = (dV/d\theta)_f = 1/[(\alpha\mu vV)/(A^2\Delta P) + (\alpha\mu L_o)/(A\Delta P)]$$

$$= 1/(2aV + b)$$

V = final volume of filtrate collected.

(b) Thorough (or through) washing, where the washing fluid is passed through the whole thickness of the filter cake. In this case the area available for washing is half the area for filtration.

$$R_w = 1/[(\alpha\mu vV)/(\tfrac{1}{2}^2A^2\Delta P) + (\alpha\mu L_o)/(\tfrac{1}{2}A\Delta P)]$$

$$= 1/(4 \times 2aV + 2b)$$

Example 5.2

A plate and frame filter press filtering a water slurry gave 8 m^3 of filtrate in 24 min and 11 m^3 in 45 min, by which time the filter was full and filtration was stopped. Estimate the time taken to wash the cake using 5 m^3 of wash water if the washing pressure is the same as that used for filtering. Assume that all operations take place at constant pressure.

For constant pressure filtration

$$(\theta/V) = aV + b$$

For the filtration (retaining minutes and m^3)

$$24/8 = 8a + b \text{ and } 45/11 = 11a + b$$

therefore

$$a = 0.3636 \text{ and } b = 0.091$$

and the final rate of filtration is given by

$$(dV/d\theta)_f = 1/(2 \times 0.3636 \times 11 + 0.091) = 1/(7.999 + 0.091)$$
$$= 1/8.09 = 0.1236 \text{ m}^3/\text{min.}$$

For simple washing the final rate of filtration is the same as the rate of washing, thus the time taken to wash with 5 m^3 of water will be 5/0.1236 = 40.4 min.

For thorough washing

$$R_w = 1/(4 \times 2aV + 2b) = 1/(4 \times 2 \times 0.3636 \times 11 + 2 \times 0.091)$$
$$= 1/(31.99 + 0.182) = 1/32.172 = 0.031 \text{ m}^3/\text{min.}$$

Therefore time taken for thorough washing with 5 m^3 of wash water will be 5/0.031 = 161 min.

Batch operation. Batch operation consists of the following stages

(a) fill the filter with suspension to vent the air

(b) filter

(c) wash the cake (if required)

(d) discharge the cake

(e) re-assemble the filter for the next batch.

For constant pressure operation with an incompressible cake, the time of filtration (θ_f) is given by

$$\theta_f = aV^2 + bV$$

if θ_w = time taken for washing, and θ_a = time for all other operations other than filtering and washing, then the output rate of the filter in terms of filtrate (V) will be

$$V/(\theta_f + \theta_w + \theta_a) = \text{say, } W \text{ m}^3/\text{s.}$$

The time of washing is a function of V, since the rate of washing the cake is proportional to the final rate of filtration ($dV/d\theta$).

Thus

$\theta_f = aV^2 + bV$ and $\theta_w = KV$ where K = a constant and the average output of the filter can be written

$W = V/(aV^2 + bV + KV + \theta_a)$

Differentiating W with respect to V and equating to zero will give an optimum volume of filtrate

$V^2 = (\theta_a/a) = (\theta_a 2A^2\Delta P)/(\alpha\mu v)$

and the optimum size of filter can be evaluated.

Example 5.3

A suspension containing 15% w/w of solids in water is to be filtered using a plate and frame filter press to give an average output of 1000 kg/h of dry solid. The plates are 1 m square, giving a filtering area per side of 0.85 m^2. The properties of the cake were found to be as follows:

specific resistance of the cake = 5×10^{12} m^{-2}; equivalent cake thickness of the cloth = 0.0075 m; volume of cake as deposited per unit volume of filtrate = 0.30; moisture content of cake = 25% w/w. Calculate the optimum number of plates and frames and the frame thickness required. The time for all other operations except filtering is 30 minutes. The cake does not require washing, is incompressible and operation of the filter takes place at 3 bar constant pressure.

For optimum operation, $V^2 = \theta_a/a$, and for constant pressure filtration

$$\theta_f = aV^2 + bV = \theta_a + bV = \theta_a + b\surd(\theta_a/a)$$

$$= \theta_a + (\alpha\mu L_o)/(A\Delta P)\surd[(\theta_a 2A^2\Delta P)/(\alpha\mu v)]$$

thus, the area required for filtration is not initially required.

$$b = (5 \times 10^{12} \times 10^{-3} \times 0.0075)/(3 \times 10^5 A) = 125/A$$

$$(\theta_a/a) = \surd[(1800 \times 2 \times A^2 \times 3 \times 10^5)/(5 \times 10^{12} \times 10^{-3} \times 0.3)]$$

$$= A\surd(0.72)$$

therefore, the optimum time of filtration is

$$\theta_f = 1800 + 125\surd(0.72) = 1800 + 106.1 = 1906 \text{ s.}$$

Total cycle time = $\theta_f + \theta_a$ = 1906 + 1800 = 3706 s.

Filtrate output rate = V/3706 m^3/s

Dry solid rate = 1000 kg/h

Wet cake rate at 25% moisture = 1000/0.75 = 1333.3 kg/h

Suspension rate at 15% dry solids = 1000/0.15 = 6666.7 kg/h

Assuming no losses, the balance on materials will be

Suspension = Wet cake + Filtrate

and 6666.7 = 1333.3 + Filtrate

Filtrate = 5333.4 kg/h.

Since the suspension consists of solid and water, the filtrate will be water, and the volumetric rate of filtrate will be 5333.4/1000 = 5.333 m^3/h. Since the optimum cycle time is 3706 s, the optimum volume of filtrate collected in this cycle time will be (5.333 x 3706)/3600 = 5.49 m^3

Now $V^2 = \theta_a/a = 1800/[(\alpha\mu v)/(2A^2\Delta P)]$

thus $A^2 = (\alpha\mu v V^2)/(2 \times 1800\Delta P)$

$= 5.49^2(5 \times 10^{12} \times 10^{-3} \times 0.3)/(2 \times 1800 \times 3 \times 10^5)$

$= 5.49^2(1.389)$

and $A = 5.49\sqrt{(1.389)} = 5.49 \times 1.179 = 6.47\ m^2$

Each plate has a filtering area of 2 x 0.85 m^2, therefore

Number of plates required = 6.47/(2 x 0.85) = 3.81

Total number of plates = 4.

The volume of cake = vV = LA where A = area of 4 plates.

LA = 0.3 x 5.49 = (2 x 0.85)L x 4

and L = (0.3 x 5.49)/(2 x 0.85 x 4) = 0.242 m.

Since the cake will build up on both sides of the plates, the thickness of the frames must cater for twice the thickness calculated above. Hence, the frame thickness must be at least 0.484 m. Thus, a total of 4 plates are required, together with a frame thickness of 0.48m.

In a real, full-scale filtration operation the time taken to reach the operating pressure for constant pressure operation is often measured in minutes, and the volume of filtrate collected during this time can be considerable.

It is common practice in such a situation to initially operate the filter under constant rate conditions until the operating pressure has been reached, and then to operate under constant pressure conditions until the filter has been filled.

Integration of the basic rate expression for the constant pressure period then yields

$$(\theta - \theta_o) = a(V^2 - V_o^2) + b(V - V_o)$$

θ_o = time for the constant rate period; V_o = volume of filtrate collected during the constant rate period; θ = total filtration time; V = total filtrate collected.

Continuous operation. The most common continuous filter is the rotary drum filter, operated either under vacuum inside the drum, or pressurised outside the drum.

The most satisfactory way of treating such filters from a design or assessment point of view is to treat the case of filtration over one revolution of the drum.

For one revolution

Area for filtration = Total surface area of the drum

Time for filtration = Time for one revolution x Immersion

Example 5.4

A rotary vacuum filter is used to filter a water slurry containing 35% w/w of solids. The drum rotates at 1 revolution in 5 min and is 1.0m diameter x1.0 m long. At any instant, 25% of the drum surface is in contact with the slurry. If filtrate is produced at the rate of 2.5 m^3/h using a vacuum of 20 inches of mercury, what will be the thickness of cake on the drum ?

If the operating conditions are changed to 25 inches of mercury vacuum and 1 revolution in 4 min, what will be the new filtrate rate ? Density of the solids = 2500 kg/m^3; voidage of the cake = 0.4. The cake is incompressible and the resistance of the filter cloth is found to be negligible.

Taking a basis of 1 revolution (5 min):

Volume of filtrate = (2.5 x 5)/ 60 = 0.2083 m^3

Solids content in cake = $(1 - \varepsilon)AL\rho_s$

ε = cake voidage; A = area of cake; L = cake thickness; ρ_s = density of the solids.

There are two possible solutions

(a) assuming that the cake voids are filled with filtrate

(b) assuming that the cake voids are filled with air.

(a) Voids filled with filtrate.

Filtrate content of the cake = $\varepsilon AL\rho$ ρ = filtrate density.

An overall mass balance will give

(solid in slurry) + (water in slurry) = (solid in cake) + (water in cake) + (filtrate)

and solid/liquid mass ratio in slurry = 0.35/0.65 = 0.5385

$= (1 - \varepsilon)AL\rho_s/(\varepsilon AL\rho + V\rho)$

so $0.5385 = (0.6 \times AL \times 2500)/(0.4 \times AL \times 1000 + 1000V)$

$215.4AL + 538.5V = 1500AL$

$AL = (538.5 \times 0.2083)/(1500 - 215.4) = 0.08732$

Area of the drum for 1 revolution = $1.0\pi \times 1.0 = 3.142\ m^2$.

Thus thickness of cake L = 0.08732/3.142 = 0.0278 m.

(b) Voids filled with air.

solid/liquid mass ratio in slurry = $(1 - \varepsilon)AL\rho_s/V\rho$

thus $0.5385 = (0.6 \times AL \times 2500)/(0.2083 \times 1000)$

$1500AL = 112.2$

and AL = 0.0748 and L = 0.0238 m.

The thickness of the cake will be between these two values, since most of the filtrate will be removed from the voids, but the cake discharged from the drum will not be completely dry. Hence the cake thickness will be between 23.8 mm and 27.8 mm.

Operation at 1 revolution in 5 minutes, 20 inches of mercury.

Filtering time/rev = 0.25 x 5 min.

For constant pressure filtration with negligible filter cloth resistance

$$\theta = [(\alpha\mu v)/(2A^2\Delta P)]\ V^2 = KV^2/\Delta P$$

$K = (\alpha\mu v)/2A^2$.

$V = 0.2083\ m^3$, and retaining inches of mercury and minutes

$(0.25 \times 5) = K(0.2083)^2/20$

K = 576.18

Operation at 1 revolution in 4 min, 25 inches of mercury

$(0.25 \times 4) = KV^2/25$ retaining inches of mercury

Since the cake is incompressible, K will be the same in both cases.

Thus $(0.25 \times 4) = 576.18V^2/25$

$V^2 = 0.04339$ and $V = 0.2083\ m^3$.

The volume of filtrate per revolution is the same in both cases, but the output rate for the second case will be

$(0.2083 \times 60)/4 = 3.1\ m^3/h$, since the time of revolution is now 4 minutes.

Compressible Filter Cakes

If the filter cake is compressible, the specific resistance of the cake will vary with pressure, as will the voidage and specific surface of the cake through the thickness. The simplest assumption in this case is to assume that the specific resistance is a function of pressure [3]

$$r = \Psi P^n \quad \text{where n is a constant.}$$

The term v, volume of cake/unit volume of filtrate, cannot now be used since the cake properties vary throughout its depth, and so using c, the mass of solids/unit volume of filtrate, the following expression is obtained for constant pressure operation

$$\theta = [(\bar{r}\mu c)/(2A^2 \Delta P)]V^2 + [(\bar{r}\mu c V_o)/(A^2 \Delta P)]V$$

$\bar{r}$ = mean specific resistance of the cake; V_o = volume of filtrate required to produce a thickness of cake equivalent to the resistance of the filter medium.

This expression is similar to those derived for incompressible filter cakes and can be used for scale-up purposes from data obtained in the laboratory, as well as for assessing the performance of an operating filter and the consequences of a change in the conditions of operation.

5.2 CENTRIFUGATION

A centrifuge is a high speed rotating cylinder which is used to enhance the gravity settling rate of either a solid in a liquid or two immiscible liquids using centrifugal forces.

The forces acting on a particle or a liquid globule in a centrifugal field are as follows

$$\text{centrifugal force} = mr\omega^2; \quad \text{gravitational force} = mg$$

m = mass of particle; r = radius from axis of rotation; ω = angular velocity; g = acceleration due to gravity.

The ratio of these forces ($r\omega^2/g$) in a typical centrifuge has a numerical value of several thousands and can be used to give a measure of the separating power of a particular machine, and can also be useful for comparing the performance of different machines.

SOLID-LIQUID SEPARATION

Two types of machine can be used for this type of separation

1. Solid wall machines where the solid particles are collected at the wall. Empty bowl (tubular) machines, machines fitted with discs and machines fitted with a scroll device to continuously remove the solids (decanter) are in current use.
2. Filtering centrifuges using a perforated basket, the pressure required for filtration being generated by centrifugal force.

Solid Bowl Machines.

The forces acting on a spherical particle in a liquid in a centrifuge are

1. A radial centrifugal force moving the particle to the wall

$$= \pi D^3(\rho_s - \rho)\omega^2 r/6$$

D = diameter of the particle; ρ_s = solid density; ρ = liquid density.

2. An accelerating force opposing particle motion

$$= \pi D^3 \rho_s (d^2r/d\theta^2)/6 \qquad \theta = \text{time}$$

3. A drag force opposing motion, the value depending on the flow regime.

Laminar region. (10^{-4}<Re'<0.2). In this region, the drag force on a spherical particle was shown by Stokes [4] to be

$$3\pi D\mu(dr/d\theta)$$

μ = liquid viscosity; Re' = particle Reynolds number = $D(dr/d\theta)\rho_s/\mu$.

Turbulent region. (500<Re'<2 x 10^5). The drag force in this region is given by

$$0.22(dr/d\theta)^2\pi D^2\rho/4$$

Since the accelerating force opposing motion contains second order terms, the numerical value compared to the other forces is negligible, and a force balance on the particle reduces to

(radial centrifugal force) = (drag force opposing motion)

In the laminar region, the approximate time for a particle to travel from a radius r_1 to a radius r_2 will be

$$\theta = 18\mu \ln(r_2/r_1)/[D^2\omega^2(\rho_s - \rho)]$$

and in the turbulent region

$$= 2(r_2^{1/2} - r_1^{1/2})/\sqrt{[3D\omega^2(\rho_s - \rho)/\rho]}$$

In operating a centrifuge for solid-liquid separation, the average residence time of the suspension in the centrifuge must be equal to the approximate time taken for a particle of a given diameter to travel from the inner liquid radius to the wall of the centrifuge, in order to collect all the particles of that diameter. If Q = volumetric feedrate of suspension and V = volume of liquid retained in the centrifuge, the average residence time = V/Q. For laminar conditions

$$V/Q = \theta = 18\mu \ln(r_2/r_1)/[D^2\omega^2(\rho_s - \rho)]$$

r_1 = inner liquid radius; r_2 = bowl radius.

If Q is adjusted to just retain a particle diameter D, then

$$Q = D^2\omega^2(\rho_s - \rho)V/[18\mu \ln(r_2/r_1)]$$

Example 5.5

A bowl centrifuge 20 cm diameter x 70 cm high is driven at 100 rev/s. With an overflow weir fixed at 5 cm radius from the central axis, the feedrate of a fruit fuice containing 1% of sediment is adjusted to give a clear overflow liquid. The smallest size of particle in the juice is 100µm. If the overflow weir is changed to 3 cm radius, what will be the change in flowrate required to give a clear overflow liquid ?

The flowrate required to just retain a particle of 100µm with an inner liquid radius of 5 cm and wall radius 10 cm will be

$$Q_1 = D^2\omega^2(\rho_s - \rho)V_1/[18\mu \ln(10/5)]$$

Similarly, when the overflow radius is changed to 3 cm

$$Q_2 = D^2\omega^2(\rho_s - \rho)V_2/[18\mu \ln(10/3)]$$

Since D, ω, ρ_s, ρ and μ remain the same

$$Q_2/Q_1 = [V_2 \ln(10/5)]/[V_1 \ln(10/3)]$$

With the overflow weir at 5 cm radius

$$V_1 = \pi \times 0.7(0.2^2 - 0.1^2)/4 = 0.0165 \text{ m}^3$$

and with the overflow weir at 3 cm, $V_2 = 0.020 \text{ m}^3$.

$$Q_2/Q_1 = [0.020 \ln(10/3)]/[0.0165 \ln(10/5)] = 0.70.$$

Thus the flowrate must be reduced to 70% of the original to maintain a clear overflow.

The free-settling velocity of a particle diameter D under

gravity is given by

$$u_o = D^2(\rho_s - \rho)g/18\mu$$

and the expression for the feedrate of suspension to a centrifuge to retain a particle of diameter D under laminar conditions can be re-written as

$$Q = u_o\omega^2 V/[g \ln(r_2/r_1)]$$

This expression for Q is only true for a single particle travelling from radius r_1 to r_2. In practice, particles of diameter D will be distributed throughout the volume of liquid in the centrifuge, and if a particle size is chosen such that 50% of such particles are retained and 50% appear in the overflow liquid, analysis of the average residence times of <u>all</u> of these particles shows that

$$Q = 2u_o\ \omega^2 V/[g \ln(r_2/r_1)] = 2u_o\Sigma$$
$$\Sigma = \omega^2 V/[\ g \ln(r_2/r_1)]$$

The value of Σ is dependent on the centrifuge dimensions and conditions of operation, and can be used to compare the performance of different centrifuges and also for the scale-up of processes. Values of Σ have been published by Trowbridge [5] and values of Q and Q/Σ (= $2u_o$) have been published for different machines [6]. Some typical values are given below.

Typical operating characteristics for centrifuges

machine	flowrate range (l/min)	Q/Σ (x 10^8 m/s)
tubular	10 - 100	5 - 35
disc	2 - 2000	8 - 50
decanter	15 - 2300	500 - 150 000

The Σ value derived above for a machine is based on a spherical particle and must be modified to take into account that the particles are not normally spherical, and also hindered settling takes place. In practice therefore, the expression for Σ must contain an efficiency term (η).

For a tubular centrifuge (laminar conditions)

$$\Sigma = \eta[\omega^2 V/g \ln(r_2/r_1)]$$

For a disc centrifuge

$$\Sigma = \eta[2\pi\omega^2(S - 1)(r_2^3 - r_1^3)/(3g \tan\delta)$$

S = number of discs; δ = disc cone half angle.

For a decanter machine

$$\Sigma = \eta 2\pi\omega^2 L(r_2^2 + 3r_1r_2 + 4r_1^2)/8g$$

L = length of liquid surface.

Efficiencies for the different machines are

tubular = 0.98; disc = 0.5; decanter = 0.60.

Typical dimensions are shown in figure 5.1.

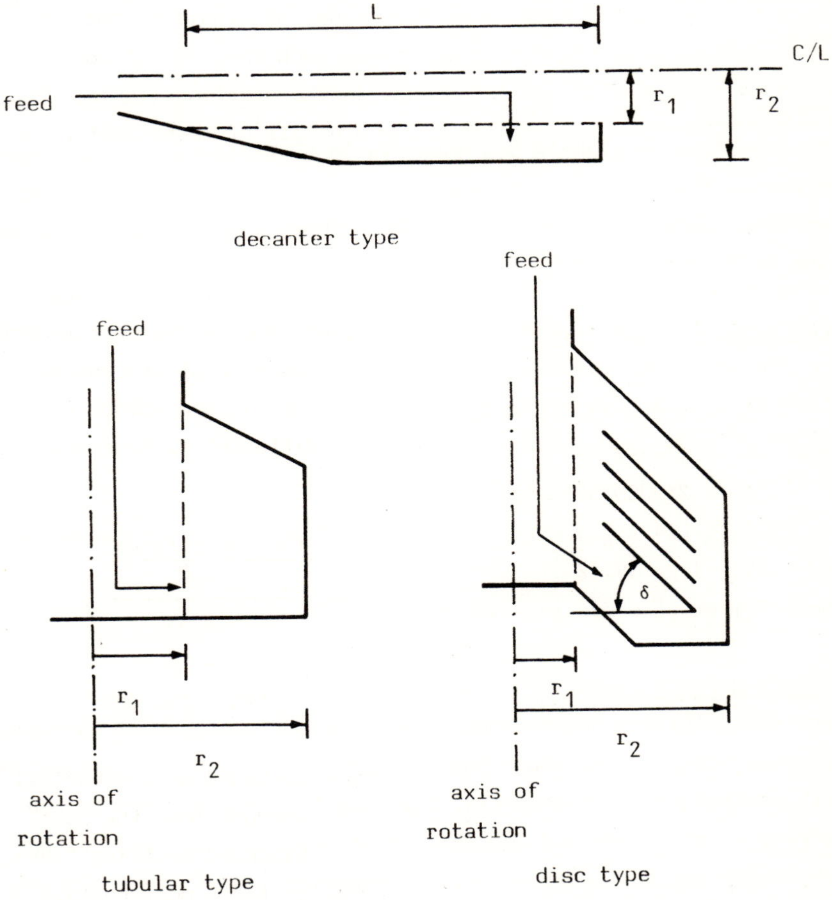

Figure 5.1 - Typical dimensions for tubular, disc and decanter centrifuges.

Example 5.6

It is required to remove particles down to and including 50% of

20μm size from 40 l/min of liquid in a centrifuge. Which type of centrifuge should be used ? Particle density = 1250 kg/m^3; liquid density = 1150 kg/m^3; liquid viscosity = 1.5 P.

$$u_o = (20 \times 10^{-6})^2(1250 - 1150) \times 9.81/(18 \times 1.5 \times 10^{-1})$$
$$= 1.4 \times 10^{-7} \text{ m/s}$$
$$Q/\Sigma = 2u_o = 2.8 \times 10^{-7} \text{ m/s.}$$

Comparing the flowrate and value of Q/Σ in the above table of operating characteristics, either a tubular or disc machine can be used.

Example 5.7

The operation in example 5.6 above is to be scaled-up to 200 l/min. Determine the number of discs required. Centrifuge speed = 5000 rev/min; outer bowl radius = 180 mm; inner liquid radius = 40 mm; cone half angle = 40°.

Throughput $Q = 200 \times 10^{-3}/60 = 3.33 \times 10^{-3}$ m^3/s

$= 2u_o\Sigma$

$2u_o = 2.8 \times 10^{-7}$ - from example 5.6

thus $\Sigma = 11\ 892$

$\omega = 2\pi 5000/60 = 523.6$ rad/s

$\Sigma = [2\pi \times 523.6^2(S - 1)(0.18^2 - 0.04^2)/(3 \times 9.81\tan 40°)]0.5$

and $(S - 1) = 59.12$

Thus, number of discs required = 61.

Filtering Centrifuges.

For free-filtering suspensions a range of perforated bowl machines are available, and although basic filtration theory can be combined with centrifugation theory [7], the resulting expressions become extremely cumbersome and complex. Most specification and design procedures are best carried out by determination of the filtering characteristics of the suspension using a pilot plant centrifugal machine, and then comparing the economics of operation against a conventional filtration plant.

LIQUID-LIQUID SEPARATION

The centrifugal pressure exerted on the wall of a centrifuge bowl due to the presence of two immiscible liquids can be shown to be

$$P_c = \tfrac{1}{2}\omega^2[\rho_1(r_2^2 - r_i^2) + \rho(r_i^2 - r_1^2)]$$

ρ_1, ρ = liquid densities ($\rho_1 > \rho$); r_i = interface radius between the two liquids.

The effect on P_c due to the head of liquid in the bowl is usually negligible compared to the centrifugal forces. For the continuous separation of two immiscible liquids using a centrifuge, the position of the overflow weir or skimmer pipe is critical, and the following relationship must be satisfied if complete separation is to be achieved [7]

$$\rho(r_i^2 - r_1^2) = \rho_1(r_i^2 - r_o^2)$$

r_o = radius of the overflow weir or skimmer pipe (see figure 5.2). Thus, r_o is a function of the densities of the two liquids and the radii r_1 (inner liquid radius - a function of the machine design and operation) and r_i (interface radius - dependent on the relative proportions of the two liquids).

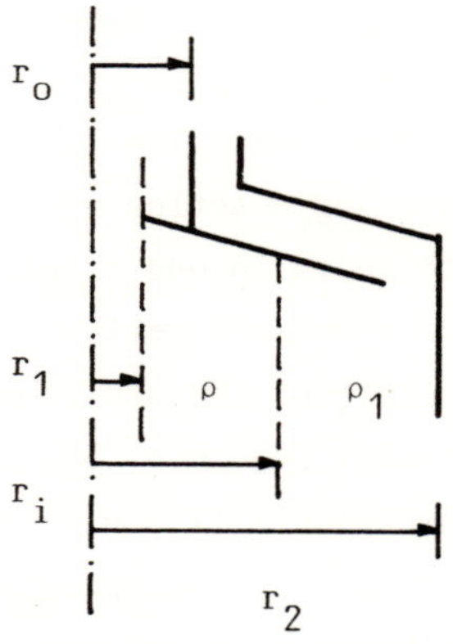

Figure 5.2 - Immiscible liquid separation. Overflow weir position.

Example 5.8

A horizontal bowl centrifuge operating at 2500 rev/min is used to separate a water-oil emulsion containing 60% v/v oil. The bowl is 300 mm diameter and 1.0 m long and is operated with half of its volume filled with emulsion. What will be the required position of the overflow weir ? Density of oil = 875 kg/m^3; density of water = 1000 kg/m^3.

Inner liquid radius r_1

Volume of liquid in bowl = $\pi r_2^2 L/2 = \pi(r_2^2 - r_1^2)L$

thus $r_1^2 = r_2^2/2$ and $r_1 = r_2/\sqrt{(2)}$ = 106 mm.

Interface radius r_i

Volume of oil = 60% of total = $0.6\pi r_2^2 L/2$

$= \pi(r_i^2 - r_1^2)L$

$r_i^2 - r_1^2 = 0.3 r_2^2$

Interface radius $r_i = \sqrt{[(0.3 \times 22\ 500) + 11\ 236]} = 134.1$ mm.

Overflow weir radius is given by

$875(r_i^2 - r_1^2) = 1000(r_i^2 - r_o^2)$

$0.875(17\ 986 - 11\ 236) = (17\ 986 - r_o^2)$

$r_o = \sqrt{(12\ 079.75)} = 109.9$ mm.

Overflow weir radius = 110 mm.

5.3 SIEVING AND CLASSIFICATION

Particle Size Distribution

The size range of a powdered or granular material can be determined by one of the following methods.

Sieve analysis. For larger particles (greater than 0.05 mm) a laboratory sieve analysis using a series of standard sieves can be carried out. Normal practice is to use a series of sieves where the aperture sizes increase as a Reynard series, i.e. by a factor of 2, $\sqrt{(2)}$ or $\sqrt[4]{(2)}$. Standard sieves are available in many National series, e.g. British Standards [8], U.S. Standard screens [9], Institute of Mining and Metallurgy (I.M.M.) [10] etc.

Microscopic analysis. A sample of material can be taken and looked at under the microscope. Using a scaled eye piece, a count of the particles of similar sizes will give an indication of the size range of the material [11].

Sedimentation analysis. For particle sizes less than 0.05 mm, the rate of sedimentation of the powder in a fluid can be used to determine the size range [12].

Permeability methods. For extremely small particle sizes, measurement of the permeability of the powder to a gas flow can be used to measure the specific surface of the powder, which can be related to the size range of the particles [13].

The results of a size analysis are usually represented graphically as a cumulative weight fraction curve where the proportion of particles is plotted against aperture size. Two methods are used for this type of plot

(a) cumulative proportion oversize versus aperture size.

(b) cumulative proportion undersize versus aperture size.

Both types of curve are shown in figure 5.3.

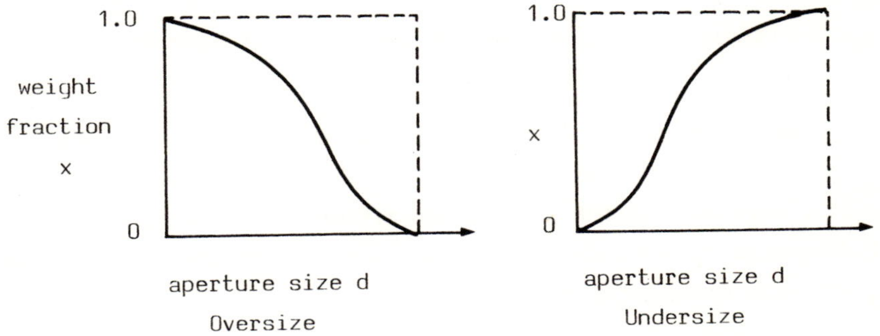

Figure 5.3 - Cumulative weight against aperture size.

These curves do not give a direct indication of the range of sizes within a material, and are used to derive additional data. For practical purposes it is convenient to express the size of a granular material using a single figure, and the simplest criterion is to use the size of aperture through which a certain proportion of material will pass. Bond [14] has suggested a figure of 80% of material passing a certain aperture, but there is no objection to using other values, e.g. 50%, 30% etc, provided that the same basis is used for comparison purposes - and the basis used is clearly stated. Simple dimensions like these can be obtained directly from the cumulative size curves.

Weight Mean Diameter (volume mean)

A typical cumulative oversize curve is shown in figure 5.4. The area enclosed by the shaded section is (d.x).

The total area enclosed by the curve and vertical axis = $\Sigma(d.x)$.

Thus the mean overall value for the diameter (d_v) is given by

$$d_v = \Sigma(d.x)/\Sigma(x) \qquad \text{...............(5.1)}$$

If the curve is a continuous mathematical function

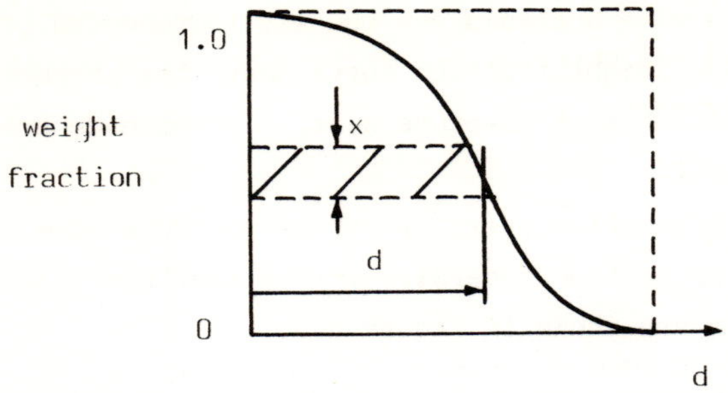

Figure 5.4 - a typical cumulative oversize curve.

Area enclosed by curve and vertical axis = $\int (d.dx)$, and

$d_v = \int (d.dx)/\int(dx)$ .... (5.2)

If a unit mass of the particulate mixture contains n_1 particles of dimension d_1 and mass fraction x_1, and n_2 particles of dimension d_2 and mass fraction x_2 etc...........

Mass fraction of particles size $d_1 = n_1 k_1 d_1^3 \rho_s = x_1$

ρ_s = solid density; k_1 = shape factor to take into account that the particles are not spherical.

Total mass fraction for all particles $= x_1 + x_2 + ... = \Sigma(x)$

$= \Sigma(n.k\ d^3 \rho_s) = 1.0$

$\Sigma(d.x) = \Sigma(n\ k\ d^4 \rho_s)$

and $d_v = \Sigma(d.x)/\Sigma(x) = \Sigma(n\ d^4)/\Sigma(n\ d^3)$ .... (5.3)

If a sieve analysis is carried out and the curve is not a simple mathematical function, the results can be used in two ways

(a) the area enclosed by a cumulative weight curve and the vertical axis can be measured to give $\Sigma(d.x)$, and equation 5.1 used to determine d_v. (This applies to both undersize and oversize curves).

(b) individual values of (d_1x_1), (d_2x_2) etc can be calculated from the sieve analysis results to give $\Sigma(d.x)$, using the average size between apertures.

If the curve is represented by a mathematical function, equation 5.2 can be used.

If the analysis was carried out by counting particles of a given size, e.g. by microscopic analysis, equation 5.3 can be used.

Example 5.9

The following data was obtained by carrying out a microscope count on a sample of a powder.

Size range (μm)	Average diameter (d)	Count (n)
0.75 - 2.25	1.50	560
2.25 - 4.50	3.375	200
4.50 - 9.00	6.75	25
9.00 - 13.5	11.25	4

Calculate the weight mean diameter of the sample.

Weight mean diameter $d_v = \Sigma(nd^4)/\Sigma(nd^3)$

d	n	nd^3	nd^4
1.50	560	1890	2835
3.375	200	7689	25949
6.75	25	7689	51898
11.25	4	5695	64072
		22963	144754

Weight mean diameter d_v = (144754)/(22963) = 6.30 μm.

Example 5.10

A sieve analysis on the powder sample analysed by microscope count in example 5.9 gave the following result

Sieve aperture (μm)	Weight retained (%)	Cumulative weight
0.75	0.0	0.0
2.25	8.2	8.2
4.50	33.5	41.7
9.00	33.5	75.2
13.50	24.8	100.0

Calculate the weight mean diameter from this data.

Weight mean diameter $d_v = \Sigma(d\ x\ )/\Sigma(x\ )$

Average diameter (d)	x	d.x
1.50	8.2	12.3
3.375	33.5	113.0
6.75	33.5	226.1
11.25	24.8	279.0
	100.0	630.4

Weight mean diameter d_v = (630.4)/(100.0) = 6.30 μm.

If the cumulative weight is plotted against aperture size, the area measured between the vertical axis and the curve is found to be 627.4, hence d_v = 6.27 μm.

Note. For a summation calculation the average diameter is used, for a graphical determination, the cumulative weight is plotted against aperture size.

Surface Mean Diameter (Sauter mean)

An alternative to a sieve or sedimentation analysis is the measurement of the permeability of the mixture, or the measurement of the surface of each mass fraction. This type of analysis is used for fine powders of sizes less than 0.01 mm. It is therefore possible to obtain a plot of cumulative surface against size, instead of a cumulative weight fraction/size curve, and a typical plot is shown in figure 5.5.

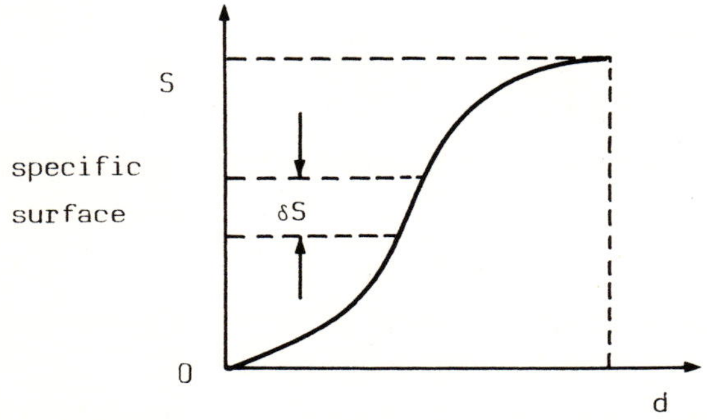

Figure 5.5 - Cumulative surface curve.

The surface mean diameter (d_s) is defined as the size of particle having the same specific surface as the mixture. For the type of curve shown in figure 5.5, the mean overall diameter is given by

$$d_s = \Sigma(d.\delta S)/\Sigma(\delta S)$$

or, for a continuous function

$$d_s = \int(d.dS)/\int(dS).$$

It can also be shown that using a particle count

$$d_s = \Sigma(n\ d^3)/\Sigma(n\ d^2)$$

and since $x_1 = (n_1 k_1 d_1^3 \rho_s)$, then using sieve analysis data

$$d_s = \Sigma(x\ )/\Sigma(x\ /d\ ).$$

Example 5.11

Using the microscope count data from example 5.9 , calculate the surface mean diameter of the powder sample.

Surface mean diameter $d_s = \Sigma(nd^3)/\Sigma(nd^2)$

Average diameter (d)	Count (n)	nd^3	nd^2
1.50	560	1890	1260
3.375	200	7689	2278
6.75	25	7689	1139
11.25	4	5695	516.3
		22963	5183.3

Surface mean diameter d_s = (22963)/(5183.3) = 4.43 μm.

Example 5.12

Using sieve analysis data from example 5.10, calculate the surface mean diameter of the powder sample.

Surface mean diameter $d_s = \Sigma(x\ )/\Sigma(x\ /d\ )$

Average diameter (d)	x	x/d
1.50	8.2	5.47
3.375	33.5	9.93
6.75	33.5	4.96
11.25	24.8	2.20
	100.0	22.56

Surface mean diameter d_s = (100.0)/(22.56) = 4.43 μm.

Linear Mean Diameter

By analogy with the weight and surface mean diameters

Linear mean diameter $d_l = \Sigma(nd^2)/\Sigma(nd)$
$= \Sigma(x\ /d\ )/\Sigma(x\ /d^2)$.

Based on the data used in examples 5.9 and 5.10, the value of the linear mean diameter d_l = 3.0 μm.

Other Dimensions

Mean volume diameter. The size each particle would have to be to make the total volume of particles the same volume as the mixture.

$$d_v^* = \sqrt[3]{[\Sigma(x_1)/\Sigma(x_1/d_1^3)]} = \sqrt[3]{[\Sigma(nd^3)/\Sigma(n)]}$$

Mean surface diameter. The size each particle would have to be to make the total surface area the same as the surface area of the mixture. (Not the specific surface).

$$d_s^* = \sqrt{[\Sigma(x_1/d_1)/\Sigma(x_1/d^3)]} = \sqrt{[\Sigma(nd^2)/\Sigma(n)]}$$

Mean linear diameter.

$$d_1^* = \Sigma(x_1/d_1^2)/\Sigma(x_1/d_1^3) = \Sigma(nd)/\Sigma(n)$$

Since all of these characteristic dimensions are widely used throughout the processing industries, it is essential to

(a) be consistent when comparing products

(b) clearly state which dimension is being used.

A summary of values obtained for the different dimensions using the data for the powder in examples 5.9 and 5.10 is given below

		Mean diameter (m)	
		Count	Sieve analysis
Weight (volume) mean	d_v	6.30	6.30
Surface (Sauter) mean	d_s	4.43	4.33
Linear mean	d_l	2.99	3.0
and for comparison purposes			
Mean volume	d_v^*	3.07	3.07
Mean surface	d_s^*	2.56	2.57
Mean linear	d_l^*	2.19	2.19

Single dimensions (d_v, d_s etc) give only one property of a solid mixture and it is often useful to plot the size frequency curve (figure 5.6), since this gives a better indication of the range of size for the particular material under consideration. This plot can be useful for deciding the size of screens to be used to give a desired product or range of products from a particular process. The frequency curve is obtained by plotting the slope of the cumulative curve against aperture size.

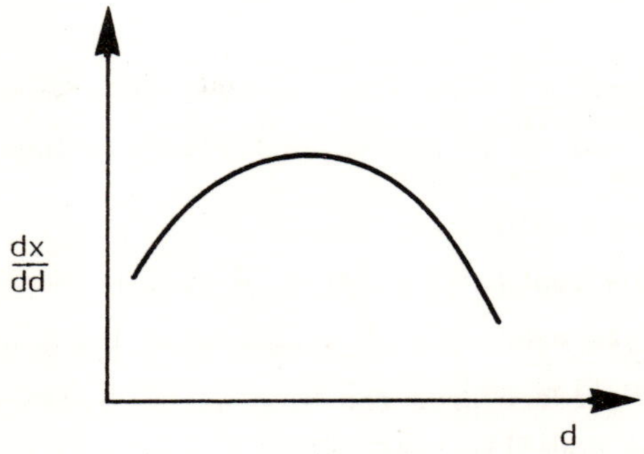

Figure 5.6 - Size frequency curve.

5.4 SOLID-LIQUID EXTRACTION

For the practical extraction of a desired solute from a solid using a liquid solvent, graphical calculation methods are more convenient than analytical methods, and this requires graphical representation of equilibrium data. Because complete separation of solid and liquid is not practical, such equilibrium data is normally based on practical experiments where the data takes into account stage efficiency.

It is common practice to represent the three-component data in other process operations using triangular diagrams, and it is possible to use this method for solid-liquid extraction [15, 16]. However, the constructions on triangular diagrams frequently lead to overcrowding in one corner of the triangle, and in many ways it is much more convenient to use rectangular co-ordinates.

Rectangular Co-ordinates

Solid-free co-ordinates are similar to solvent-free diagrams used in liquid-liquid extraction [7,17].

If A = solvent; B = insoluble carrier solid; C = desired solute

N = solid content = weight ratio B/(A + C)

Y = solute content = weight ratio C/(A + C) in the OVERFLOW

X = solute content = weight ratio C/(A + C) in the UNDERFLOW

The OVERFLOW is the clear decanted liquid after contact, the UNDERFLOW is the separated solid or slurry after contact.

The types of equilibrium curves found in practice depend on the properties of the particular solid/solvent/solute system.

System type 1. Figure 5.7 represents a system where

(a) the overflow solution and the solution associated with the solid have the same composition

(b) the solute is soluble in all proportions with the solvent

(c) the overflow contains no solid, either suspended or dissolved.

The tie-lines are used to link equilibrium solution concentrations in the overflow and underflow — for this type of system, vertical lines. The distribution diagram shown in figure 5.7 is a plot of tie-line values and is useful for interpolation purposes

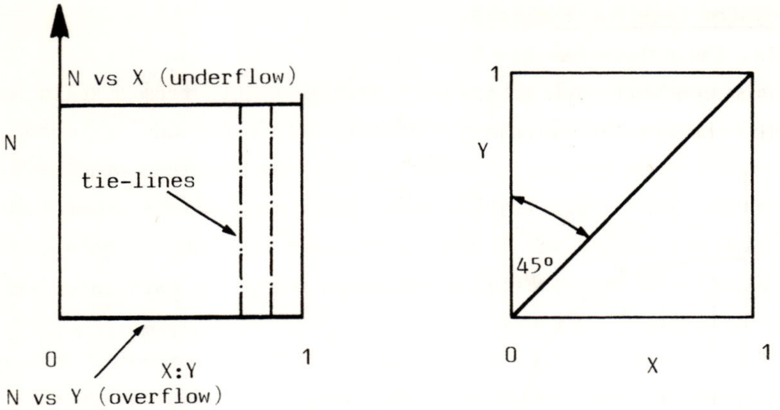

Equilibrium diagram Distribution diagram

Figure 5.7 - Equilibrium diagram - type 1 system.

(for this type of system where Y = X, the distribution diagram is not necessary).

System type 2. Figure 5.8 represents a system where

(a) the solute is soluble in all proportions with the solvent

(b) the solid is either partially soluble in the solvent, or some solid is withdrawn with the overflow liquid ($N \neq 0$)

(c) either the solute is soluble in the solid a small extent, or insufficient time is allowed for complete equilibrium, or preferential adsorption of solute occurs in the solid (tie-lines are skew).

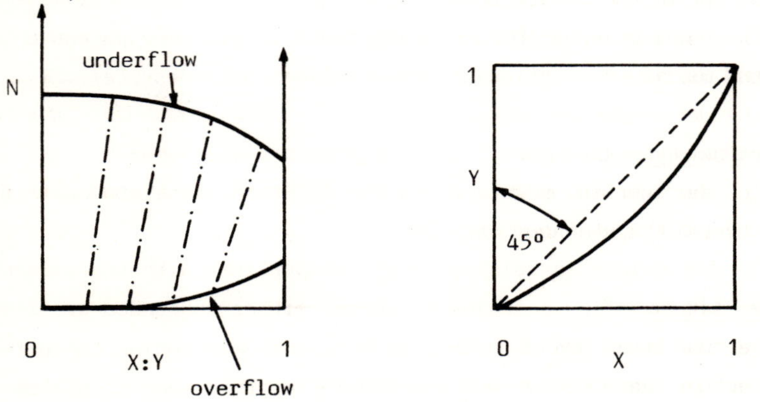

Equilibrium diagram Distribution diagram

Figure 5.8 - Equilibrium diagram - type 2 system.

System type 3. Figure 5.9 represents a system where

(a) the solute has limited solubility in the solvent (Y_s = maximum concentration in the overflow liquid)

(b) the overflow liquid contains no solid.

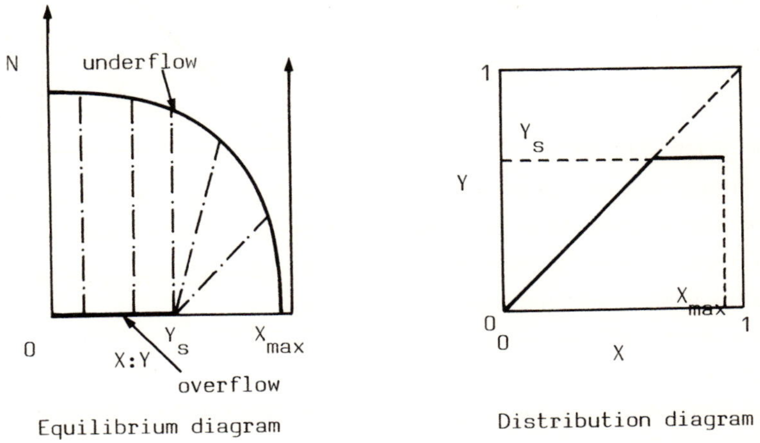

Figure 5.9 - Equilibrium diagram - type 3 system.

The basic extraction operation is as follows

(1) Mix the solid feed with the solvent, providing as much area as possible for contact, e.g. using crushed solid.

(2) Allow an adequate time to closely approach equilibrium.

(3) Separate the solid (underflow) from the liquid (overflow)

Thus, the basic system design can be summarised as

(a) perform materials balances to determine the characteristics of the mixture

(b) determine the equilibrium or practical concentrations of the overflow and underflow

(c) perform materials balances to determine the quantities of the overflow and underflow.

Nomenclature

F = actual quantity of feed solid; S = actual quantity of solvent;

E = actual quantity of overflow; R = actual quantity of underflow;

M = total mixture = F + S.

Solid-free quantities. F' = solid-free feed; S' = solid-free solvent; E' = solid-free overflow; R' = solid-free underflow; M' = solid-free mixture = $F' + S'$.

The relationships between the solid-free quantities and the real, actual quantities are

$$F = F'(1 + N_F); \quad S = S'(1 + N_S); \quad E = E'(1 + N_E); \quad R = R'(1 + N_R).$$

Single Stage Contact

In a single contact operation, the feed solids are mixed with the solvent for a given time to allow equilibrium to be approached, and the mixture is then allowed to settle. After settling, the overflow liquid is separated from the underflow solids.

The overall solid-free balance is

$$F' + S' = M' = E' + R'$$

The solid-free solute balance

$$F'X_F + S'Y_S = M'X_M = E'Y_1 + R'X_1$$

The solids balance (solid-free basis)

$$F'N_F + S'N_S = M'N_M = E'N_E + R'N_R$$

From the above balances

$$X_M = (F'X_F + S'Y_S)/(F' + S') \text{ and } N_M = (F'N_F + S'N_S)/(F' + S')$$

(X_M, N_M) being the co-ordinates of a point representing M' on an equilibrium diagram plot of N against X, Y. (See example 5.13 below).

It can also be derived from the above balances that

$$(F'/S') = \text{weight ratio } F' : S' = (N_M - N_S)/(N_F - N_M)$$

and the point M' lies on the straight line $F'S'$ such that the weight ratio $F' : S'$ = length ratio $(S'M'/M'F')$. This is known as the inverse lever rule [7].

Example 5.13

A ground fishmeal containing 40% w/w of oil is to be contacted in a single stage with benzene to extract the oil. Experimental data for the extraction is given in table 5.1. If 1000 kg of fishmeal is contacted with 750 kg of benzene, (a) what will be the quantities and compositions of the overflow liquid and underflow solid ? (b)what will be the percentage of oil extracted ?

Table 5.1

Experimental Equilibrium Data - Fishmeal and Benzene.

Concentration of oil in solution (X,Y)	kg solution retained per kg solid	Underflow solid content (N)
0.0	0.5	2.0
0.1	0.505	1.980
0.2	0.515	1.942
0.3	0.53	1.887
0.4	0.55	1.818
0.5	0.571	1.751
0.6	0.595	1.681
0.7	0.62	1.613
0.8	0.65	1.538
0.9	0.68	1.471
0.99	0.714	1.40

X = Y, the tie-lines are vertical.

The equilibrium diagram is shown in figure 5.10.

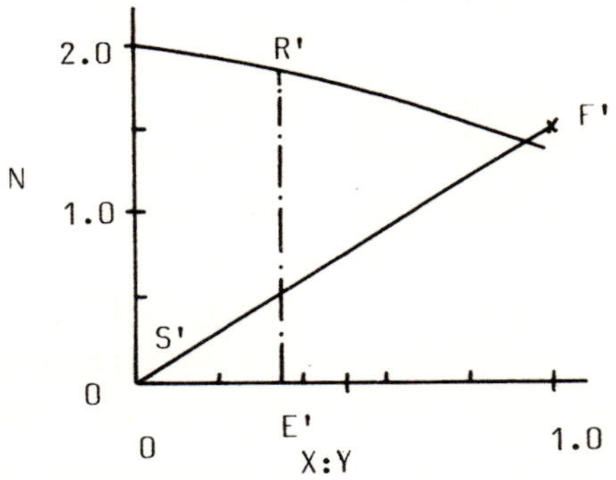

Figure 5.10 - Experimental data for fishmeal/oil/benzene. Single stage extraction, example 5.13.

(a) The fishmeal contains 40% w/w of oil, thus

$X_F = (0.4/0.4) = 1.0$; $N_F = (0.6/0.4) = 1.50$ - locating the point representing F' on figure 5.10.

$F' = 1000 \times 0.4 = 400$ kg (solid-free).

Solute balance

$$F'X_F + S'Y_S = M'X_M = E'Y_1 + R'X_1$$

$$X_M = (400 \times 1.0 + 0.0)/(400 + 750) = 0.3478 \quad Y_S = 0.0$$

Solid balance

$$F'N_F + S'N_S = M'N_M = E'N_E + R'N_R$$

$N_M = (400 \times 1.5 + 0.0)/(400 + 750) = 0.5217 \quad N_S = 0.0$

These two values (X_M, N_M) locate point M' on figure 5.10 on the straight line F'S', and since the tie-lines are vertical

$X_1 = Y_1 = X_M = 0.3478$

The intersection of the tie-line through X_M with the overflow line ($N_E = 0.0$) gives $N_E = 0.0$.

The intersection of the tie-line through X_M with the underflow line gives $N_R = 1.852$.

From the solid balance

$M'N_M = E'N_E + R'N_R$

thus $(400 + 750)0.5217 = 0.0 + 1.852R'$

and $R' = 323.95$ kg (solid-free)

$R = 323.95(1 + 1.852) = 924$ kg (actual quantity).

now $E' = M' - R' = 1150 - 323.95 = 826.05$ kg

and $E = 826.05(1 + 0) = 826$ kg (actual)

Overflow liquid = 826 kg at 34.78% oil concentration; underflow solid = 924 kg.

(b) Original oil in the feed = 1000 x 0.4 = 400 kg

Oil concentration in overflow = 34.78%

Quantity of overflow liquid = 826 kg

Oil in overflow = 826 x 0.3478 = 287.3 kg.

Percentage of oil extracted = (287.3/400) x 100 = 71.8%.

Multiple-stage Cross Contact

In practice it is found to be more efficient to contact the feed solid in successive stages with a small amount of solvent at each stage than to use a single contact process using the total amount of solvent.

The calculations involved for each stage are essentially the same as for a single contact outlined above, but the feed for successive stages is the underflow solid from the previous stage.

First stage. The solid-free balances are as follows

Overall $F' + S'_1 = M'_1 = E'_1 + R'_1$

Solid $F'N'_F + S'_1N_S = M'_1N_{M1} = E'_1N_{E1} + R'_1N_{R1}$

Solute $F'X_F + S'_1Y_S = M'_1X_{M1} = E'_1Y_1 + R'_1X_1$

Calculation of N_{M1} and X_{M1} will locate the point representing M'_1 on the straight line $F'S'_1$, and the tie-line through M'_1 will give the concentrations of the overflow and underflow (Y_1 and X_1) for the first stage.

Second stage. Since the underflow from the first stage becomes the feed for the second stage, the solid-free balances are

Overall $R'_1 + S'_2 = M'_2 = E'_2 + R'_2$

Solid $R'_1N_{R1} + S'_2N_S = M'_2N_{M2} = E'_2N_{E2} + R'_2N_{R2}$

Solute $R'_1X_1 + S'_2Y_S = M'_2X_{M2} = E'_2Y_2 + R'_2X_2$

The point representing M'_2 lies on the straight line $R'_1S'_2$, and can be located by calculation of N_{M2} and X_{M2}. The tie-line passing through M'_2 will give the concentrations of the overflow and underflow (Y_2 and X_2) for the second stage.

Subsequent stages. Each of the subsequent stages takes the underflow from the previous stage. Normal practice is to bulk together all the overflow liquids from all stages for a common recovery of the solvent.

Example 5.14

1000 kg of the oil-bearing fishmeal in example 5.13 is to be contacted in two stages, using 400 kg of benzene in the first stage, 350 kg of benzene in the second stage. Calculate (a) the quantities of underflow and overflow for each stage (b) the percentage of oil extracted.

(a) First stage. $X_F = 1.0$; $F' = 400$ kg; $S'_1 = 400$ kg; $N_F = 1.50$; $N_S = 0.0$; $Y_S = 0.0$.

$$X_{M1} = (F'X_F + S'_1Y_S)/(F' + S'_1) = (400 \times 1 + 0.0)/800 = 0.5$$

$$N_{M1} = (F'N_F + S'_1N_S)/(F' + S'_1) = (400 \times 1.5 + 0)/800 = 0.75$$

Since the tie-lines are vertical

$X_{M1} = Y_1 = X_1$, and from the construction shown on figure 5.11, $N_{E1} = 0.0$ and $N_{R1} = 1.751$.

From the solid balance

$$M'_1N_{M1} = E'_1N_{E1} + R'_1N_{R1}$$

thus $(800 \times 0.75) = 0.0 + 1.751R'_1$

$R'_1 = 600/1.751 = 342.66$ kg (solid-free)

$R_1 = 342.66(1 + 1.751) = 942.66$ kg (actual).

and $E'_1 = (800 - 342.66) = 457.34$ kg $= E_1$ (since $N_{E1} = 0.0$).

<u>Second stage</u>. $R_1' = 342.66$ kg; $X_1 = 0.5$; $S_2' = 350$ kg; $N_S = 0.0$; $Y_S = 0.0$.

$X_{M2} = (R_1'X_1 + S_2'Y_S)/(R_1' + S_2') = (342.66 \times 0.5)/(692.66) = 0.247$

$N_{M2} = (342.66 \times 1.751)/(692.66) = 0.866.$

From figure 5.11, $Y_2 = X_2 = X_{M2} = 0.247$

$N_{E2} = 0.0$, $N_{R2} = 1.913$.

From the solid balance

$M_2'N_{M2} = E_2'N_{E2} + R_2'N_{R2}$

thus $(692.66 \times 0.866) = 0.0 + 1.913R_2'$

$R_2' = 313.56$ kg (solid-free)

$R_2 = 313.56(1 + 1.913) = 913.4$ kg (actual)

and $E_2' = (692.66 - 313.56) = 379.1$ kg $= E_2$ (since $N_{E2} = 0.0$).

Summary.

First stage - overflow = 457 kg. ; underflow = 942 kg.

Second stage - overflow = 379 kg. ; underflow = 913 kg.

(b) Original oil in feed = 400 kg.

Oil in first stage overflow = 457.34 x 0.5 = 228.67 kg.

Oil in second stage overflow = 379.1 x 0.247 = 93.64 kg.

Total oil extracted in two stages = 323.2 kg

Percentage extracted = (323.2/400) x 100 = 80.6%

(This compares with 71.8% in a single stage using the same total solvent - example 5.13).

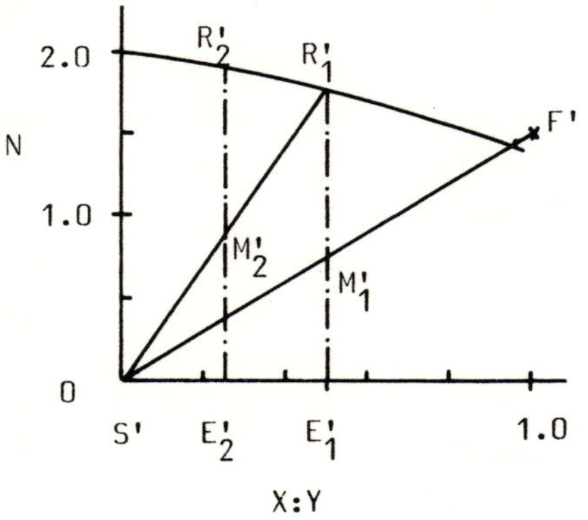

Figure 5.11 - Construction for multiple-stage cross contact, example 5.14.

Continuous Multiple-stage Counter Current Contact

The flow diagram for the overall process of n stages shown in figure 5.12.

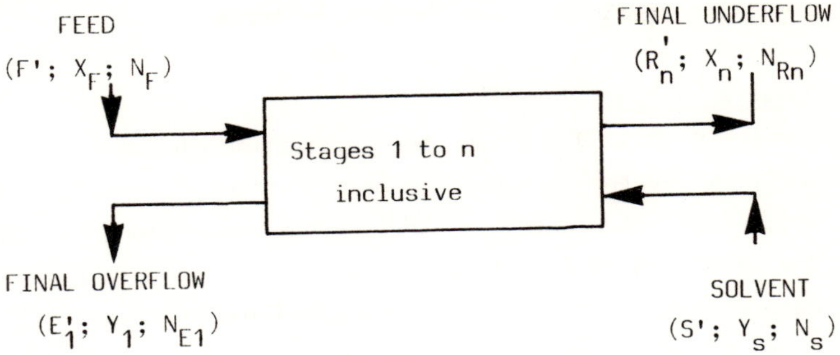

Figure 5.12 - Flow diagram for n stages of continuous contact.

The overall flow balance on a solid-free basis for n stages is as follows

$F' + S' = M'_0 = E'_1 + R'_n$, where the quantities are now flow rates.

Rearranging gives

$F' - E'_1 = R'_n - S'$ = some flowrate, say P'.

The inverse lever rule tells us that the point representing P' must lie on the straight line $F'E'_1$ extended and on the straight line R'_nS' extended. Thus, point P' is obtained by extending $F'E'_1$ and R'_nS' to the point of intersection. This point is known as the operating point, shown on figure 5.13.

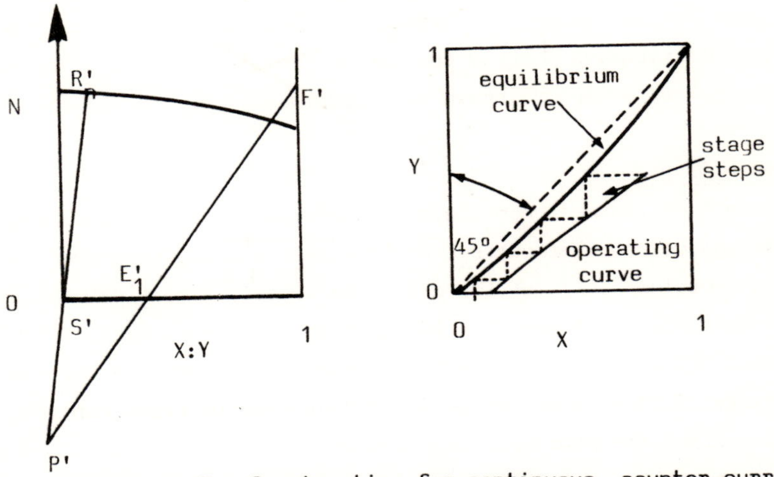

Figure 5.13 - Construction for continuous, counter-current contact.

Overall Balance, Stage 2 to n inclusive:

$$R'_1 + S' = E'_2 + R'_n$$

and $R'_n - S' = R'_1 - E'_2 = P'$

thus, line $R'_1E'_2$ extended passes through point P'

Calculation Steps.

1. Locate F', S', E'_1 & R'_n on the equilibrium diagram
2. Locate M'_o at the intersection of F'S' and $R'_nE'_1$ and X_M or N_M will allow calculation of the solvent requirement to achieve the specified E'_1 and R'_n compositions
3. Locate the operating point P' by extending R'_nS' and $F'E'_1$
4. Since R'_1 and E'_1 are in equilibrium (or at experimental values), the tie-line from E'_1 will locate R'_1
5. Draw $P'R'_1$ and the intersection with the overflow curve gives E'_2
6. The tie-line from E'_2 locates R'_2
7. Continue as from 5 above until R'_n is reached.

The number of TIE-LINES gives the number of stages required.
In order to obtain a realistic intersection point for P' on reasonable sized paper, there is no objection to distorting the scales of the plot, since the inverse lever rule is applicable even with widely varying scales.

If a large number of stages are required, it is more convenient to use the distribution diagram. Draw random lines from P' to intersect the overflow and underflow curves, and plot these values of X and Y on the distribution diagram to give an Operating Curve. Step off the number of stages as shown in figure 5.13.

Example 5.15

1000 kg/h of fishmeal containing 45% w/w oil is to be extracted in a continuous, counter-current contact system using recovered benzene which contains 5% w/w of the oil. The process specifications are such that the final overflow liquid from the first stage is to contain 45% w/w oil, and the final underflow solid from the last stage is to contain not more than 10% oil (solid-free basis). (a) What solvent feedrate will be required to meet these specifications ? (b) What will be the flowrates of the final overflow and underflow streams ? (c) How many stages will be required to meet the specifications ? (d) What will be the overall recovery of the oil ? (The experimental data given in example 5.13 above is applicable to this system).

The experimental data is plotted on figure 5.14.

(a) $F = 1000$ kg/h at 45% oil; $F' = 450$ kg/h (solid free)

$X_F = 0.45/0.45 = 1.0$; $N_F = 0.55/0.45 = 1.222$

$X_S = 0.05/1.0 = 0.05$; $N_S = 0.0$

The above co-ordinates locate points F' and S'

$X_n = 0.10$; $Y_1 = 0.45$ (locates points R'_n and E'_1)

At the intersection of lines F'S' and $E'_1R'_n$ from figure 5.14

$X_M = 0.375$; $N_M = 0.415$

A solid balance (overall) gives

$$F'N_F + S'N_S = (F' + S')N_M$$

$$(450 \times 1.222) = (450 + S')0.415$$

and $S' = 875$ kg/h.

Since $N_S = 0.0$, this will be the actual solvent rate required.

Required solvent feed rate = 875 kg/h.

(b) At point R'_n $X_n = 0.10$; $N_{Rn} = 1.98$ (figure 5.14)

At point E'_1 $Y_1 = 0.45$; $N_{E1} = 0.0$

$$(F' + S')N_M = E'_1N_{E1} + R'_nN_{Rn}$$

$$(450 + 875)0.415 = 0.0 + R'_n \times 1.98$$

$$R'_n = (1325 \times 0.415)/1.98 = 277.7 \text{ kg/h}$$

$$R_n = 277.7(1 + 1.98) = 827.6 \text{ kg/h (actual)}$$

and $E'_1 = (1325 - 277.7) = 1047.3$ kg/h

$= E_1$ (since $N_{E1} = 0.0$)

Thus, final overflow rate = 1047 kg/h

final underflow rate = 828 kg/h

(c) The intersection of lines $F'E'_1$ and R'_nS' extended meet at point P' (see figure 5.14).

Since the tie-lines are vertical, $X_1 = Y_1$ locating R'_1.

The line drawn from P' to R'_1 intersects the overflow curve (N = 0) at E'_2, and since $X_2 = Y_2$ the vertical tie-line locates R'_2.

Similarly, the line $P'R'_2$ is used to locate E'_3 and thus R'_3.

Since X_3 is smaller than the specified X_n, three stages will more than meet the requirements.

Thus, number of stages required = 3.

(d) Total amount of oil available for recovery is the sum of the oil content of the feed and the oil in the solvent.

Total available = (450 x 1.0) + (875 x 0.05)

= 493.75 kg/h.

Oil in Final Overflow = 45%.

Oil recovered = (1047.3 x 0.45) = 471.285 kg/h.

Recovery = (471.285/493.75) = 0.9545 i.e. 95.5%

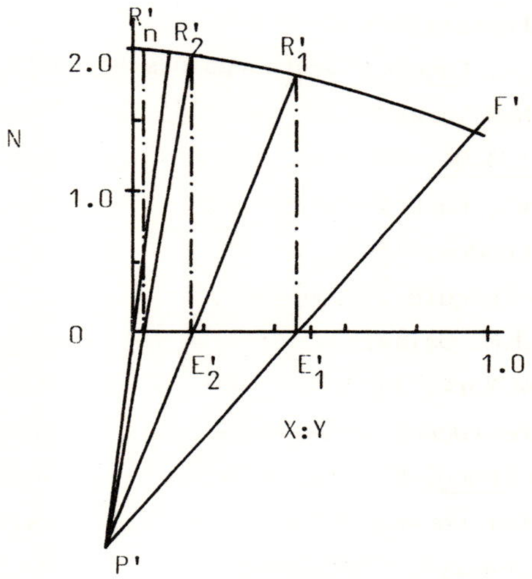

Figure 5.14 - Continuous, counter-current contact, example 5.15.

REFERENCES

1. H.P.G. Darcy, Les fontaines publique de la ville de Dijon, (Victor Dalamont, 1856)
2. P.C. Carman, 'Flow through granular beds', Trans.Instn chem. Engrs, 15 (1937) 150
3. H.P. Grace, 'Resistance and compressibility of filter cakes' Chem.Engng Prog., 49 (1953) 303
4. G.G. Stokes, 'On the effects of the internal friction of fluids on the motion pendulum', Trans.Camb.Phil.Soc., 9 (1851) 8

5. M.E.O'K. Trowbridge, 'Problems in the scaling-up of centrifugal separation equipment', Chem.Engr London, 162 (1962 - Aug) A73
6. B.G. Morris, 'Applications and selection of centrifuges', Br.chem.Engng, 11 (1966) 846
7. J.M. Coulson and J.F. Richardson, Chemical Engineering, Vol.2, (Pergamon, Oxford, 1978)
8. Br.Stand.Specif.410, Specification for test sieves', (Br. Stands Instn, London, 1976)
9. A.S.T.M.Stand.E11, 'U.S. Standard woven wire sieves', (Am.Soc. Testing and Materials, 1970)
10. J.H. Perry (Ed.), Chemical Engineers' Handbook, 3rd Ed., (McGraw-Hill, New York, 1950)
11. Br.Stand.Specif.3406, 'Methods for determination of particle size of powders', Part 4 - Optical microscope methods, (Br. Stands Instn, London, 1963)
12. ibid, Part 2 - 'Liquid sedimentation methods', (1963)
13. C. Orr Jr and J.M. Dalavalle, Fine particle measurement, (Macmillan, New York, 1959)
14. F.C. Bond, 'Some recent advances in grinding theory and practice', Br.chem.Engng, 8 (1963) 631
15. W.J. George, 'For non-equilibrium leaching - calculate number of steps graphically', Chem.Engng, 66(3) (1959) 111
16. E. Sattler-Dornbacher, 'Berechnung der Gegenstromextractionen fest/flussig', Chem.Ing.Tech., 30 (1958) 14
17. R.E. Treybal, Mass Transfer Operations, 3rd Ed., (McGraw-Hill, New York, 1980)

6 MIXING

6.1 LIQUID/LIQUID MIXING

When two liquids are mixed in a stirred vessel there are two important factors to be considered in evaluating the mixer performance (a) power consumption and (b) the degree of mixing obtained.

Power consumption

In general this depends on the flow conditions in the mixer, as

Type of Impeller	D_t/D_i	Z_1/D_i	Z_t/D_i	Baffles No.	Baffles w/D_t	No.	Ref.	Type of Impeller	D_t/D_i	Z_1/D_i	Z_t/D_i	Baffles No.	Baffles w/D_t	No.	Ref.
Turbine with 6 flat blades. 0.25D_i, 0.2D_i	3	2.7-3.9	0.75-1.3	4	0.17	1	7	Paddle with 2 blades. 0.25D_i	4.35	4.3	0.29	3	0.11	8	3
Same as No. 1.	3	2.7-3.9	0.75-1.3	4	0.10	2	7	Paddle with 4 blades. See No. 8.	3	3	0.5	0		16	2
Same as No. 1.	3	2.7-3.9	0.75-1.3	4	0.04	4	7	Paddle with 2 blades. See No. 8.	3	3.2	0.33	0		20	4
Same as No. 1. $a = 1$, $b = 40$	3	2.7-3.9	0.75-1.3	0		14*	7	Paddle with 2 blades. See No. 8.	3	2.7-3.9	0.75-1.3	4	0.10	10	7
Turbine with 6 curved blades. Blade sizes same as No. 1.	3	2.7-3.9	0.75-1.3	4	0.10	3	7	Paddle with 2 blades. See No. 8. Blade width = 0.13D_i.	1.1	0.5	0.19	0		29	10
Turbine with 6 arrowhead blades. Blade sizes same as No. 1.	3	2.7-3.9	0.75-1.3	4	0.10	5	7	Paddle with 2 blades. See No. 8. Blade width = 0.17D_i.	1.1	0.4	0.10	0		29	10
Radial turbine with deflector ring.				0		7	9	Marine propeller with 3 blades. Pitch = 2D_i	3	2.7-3.9	0.75-1.3	4	0.10	15	7
Shrouded turbine with 6 blades, 20-blade deflector ring.	2.4	0.74	0.9	0		11	6	Same as No 15. $a = 1.7$, $b = 18$.	3.3	2.7-3.9	0.75-1.3	0		21*	7
Similar to No. 11, but not identical.	3	2.7-3.9	0.75-1.3	0		12	7	Same as No. 15, but pitch = 1.33D_i	16			3	0.06	18	5
Same as No. 12, but no deflector ring.	3	2.7-3.9	0.75-1.3	4	0.10	13	7	Same as No. 15, but pitch = 1.09D_i.	9.6			3	0.06	23	5
Axial turbine with 8 blades at 45° angle. See No. 17.	3	2.7-3.9	0.75-1.3	4	0.10	9	7	Same as No. 15, but pitch = 1.05D_i, $a = 2.3$, $b = 18$.	2.7	2.7-3.9	0.75-1.3	0		27*	7
Axial turbine with 4 blades at 60° angle. 0.25D_i	3	3	0.50	0		17	2	Same as No. 15, but pitch = 1.04D_i, $a = 0$, $b = 18$.	4.5	2.7-3.9	0.75-1.3	0		25*	7
Axial turbine with 4 blades at 45° angle. See No. 17.	5.2	5.2	0.87	0		19	2	Same as No. 15, but pitch = D_i.	3	2.7-3.9	0.75-1.3	4	0.10	24	7
Same as No 19.	2.4-3.0	2.4-3.0	0.4-0.5	0		22	2	Same as No. 15, but pitch = D_i. $a = 2.1$, $b = 18$.	3	2.7-3.9	0.75-1.3	0		26*	7
Disk with 16 vanes 0.1D_i, 0.35D_i	2.5	2.5	0.75	4	0.25	6	1	Same as No. 15, but pitch = D_i.	3.8	3.5	1.0	0		28	8

D_i = diameter of impeller, D_t = diameter of tank, n = revolutions per second, w = width of baffle, Z_1 = elevation of impeller above tank bottom, Z_t = height of liquid in tank.

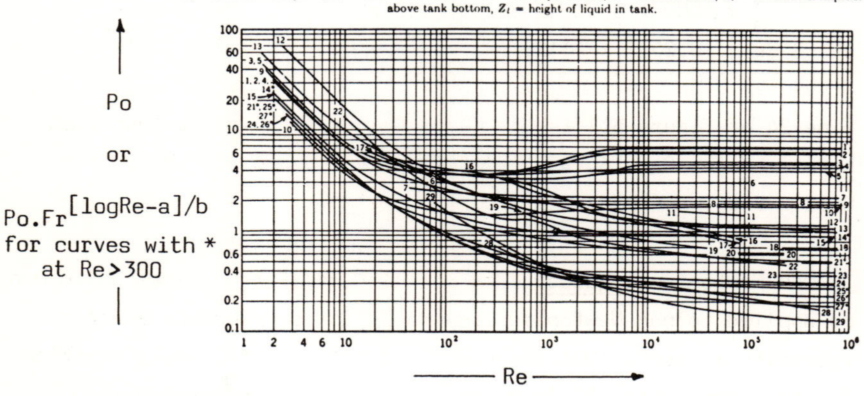

Figure 6.1: Power consumption in liquid mixing (from [4])

defined by a Reynolds Number and a Froude Number, the latter being a measure of the extent of vortex production which is usually only significant in unbaffled vessels. Figure 6.1 shows typical relationships between power consumption (the Power Number Po = $P/\rho_m N^3 d_a^5$), Reynolds Number (Re = $d_a^2 N \rho_m/\mu_m$) and Froude Number (Fr = $d_a N^2/g$) for a number of mixer configurations. In these dimensionless groups P is the power transmitted via the agitator, N and d_a are the agitator speed and diameter, ρ_m and μ_m are the mixture density and viscosity. It has been proposed [1] [2] that $\rho_m = (v_1\rho_1 + v_2\rho_2)$ where v_1 and v_2 are the volume fractions of liquids 1, the continuous phase, and 2, the dispersed phase in the mixture. For an unbaffled vessel $\mu_m = \mu_1^{v1}\mu_2^{v2}$ [1] and for a baffled vessel $\mu_m = (\mu_1/v_1)\{1 + [1.5\mu_2 v_2/(\mu_1 + \mu_2)]\}$ [3].

Example 6.1

Calculate the required wattage of the agitator motor for the following mixing operation. The mixer has a propeller stirrer of pitch and diameter 25cm in a baffled tank of 1m diameter, the stirrer running at 1150 r.p.m. The liquids being mixed have viscosities of 0.03 and 0.01kg/ms and densities of 850 and 1000kg/m^3 in a volume ratio of 1:3. The liquid of larger volume fraction forms the continuous phase.

$$\mu_m = (0.01/0.75)\{1 + [1.5 \times 0.03 \times 0.25/(0.01 + 0.03)]\} = 0.01708\text{kg/ms}$$

$$\rho_m = 1000 \times 0.75 + 850 \times 0.25 = 962.5\text{kg/m}^3$$

$$Re = (0.25)^2 \times (1150/60) \times 962.5/0.01708 = 6.75 \times 10^4$$

Line 24 of figure 6.1 is the nearest configuration to that used here, and for Re = 6.75×10^4 this gives Po = 0.32. Hence P = $0.32 \times 962.5 \times (1150/60)^3 \times 0.25^5$ = 2118W = 2.1kW

Non-Newtonian liquids. The correlations for Po given above can be applied to non-Newtonian liquids if the appropriate apparent viscosity is used. Uhl and Gray [5] give the following equations for the shear rate ($\dot{\gamma}$) in propeller or turbine stirred mixing vessels: for dilatant fluids $\dot{\gamma} = 38N(d_a/d_t)^{0.5}$ and for pseudoplastic and Bingham plastic fluids $\dot{\gamma} = 10N$ (d_t is the vessel diameter). The apparent viscosity is then obtained from the flow curve (viscosity versus shear rate) for the liquid.

Example 6.2

The continuous phase liquid in example 6.1 is a pseudoplastic material of density $1000 kg/m^3$ showing the following shear stress (R): shear rate ($\dot{\gamma}$) relationship: $R = 4(\dot{\gamma})^{0.5}$. Calculate the power consumption in the mixing system of example 6.1.

For a pseudoplastic fluid $\dot{\gamma} = 10N$. Thus $\dot{\gamma} = 10 \times 1150/60 = 192 s^{-1}$. Hence $R = 4 \times 192^{0.5} = 55.4$. Hence apparent viscosity $= R/\dot{\gamma} = 0.2886$ kg/ms. Thus, $\mu_m = (0.2886/0.75)\{1 + [1.5 \times 0.003 \times 0.25/(0.2886 + 0.003)]\} = 0.386$kg/ms. $\rho_m = 962.5$ kg/m^3 (as before). Thus $Re = 0.25^2 \times (1150/60) \times 962.5/0.386 = 2987$ and $Po = 0.36$ (from figure 6.1, line 24). From this $P = 0.36 \times 962.5 \times (1150/60)^3 \times 0.25^5 =$ 2.4kW.

Rate of mixing

An analysis of the mixing progress, i.e. the rate at which the degree of 'mixedness' changes with time, requires a measure of the extent of mixing. The most common method of describing this is in terms of the variability of one component in samples taken from different locations in the mixer. The standard deviation of the fractional volume of one component in samples taken from the initially completely unmixed two-component system (σ_o) will be $\{v_1(1-v_1)\}^{0.5}$ where v_1 is the average fractional volume of that component in the mixture. If a further n samples are taken after a certain mixing time, θ_m, then the standard deviation of the fractional volume of one component will be, by definition, $\sigma_m = \{[1/(n-1)] \sum_{i=1}^{n} (v_i-v_1)^2\}^{0.5}$, where v_i is the fractional volume in sample i.

A mixing index (M) can be defined as $M = (\sigma_m^2-\sigma_\infty^2)/(\sigma_o^2-\sigma_\infty^2)$, where σ_∞ is the standard deviation of a 'perfectly mixed' sample. Although complete uniformity (i.e. $\sigma_\infty = o$) cannot be reached, in practice for liquids a value of σ_∞ very close to zero will be achievable after a reasonable time. Generally it is found that M =

$\exp(-k\theta_m)$, where k is a mixing rate constant dependent on the characteristics of the mixer and the liquids. The effect of mixer characteristics on k can be approximately established from $k \propto d_a^3 N/d_t^2 H$ where H is the liquid height in the mixing vessel [6]. The effect of liquid properties on k can only be satisfactorily established by experiment; Leniger and Beverloo [7] suggest $k \propto D_v$, where D_v is the diffusivity of one component of the mix into the second.

Example 6.3

The following data were obtained for samples taken from a mixer blending equal quantities of two liquid components of a soup mix. Calculate (a) the time for 95% of the samples to fall within ±4% of the mean fraction of one ingredient, (b) this time if the mixer speed were increased by 50% and (c) this time if the batch size in the mixer were reduced by 20%.

Time(min) :	1	2	3	4	5	6
σ_m :	0.303	0.236	0.151	0.0913	0.0582	0.0410

Since the components are in equal quantity, $v_1 = 0.5$. Therefore $\sigma_o = (0.5 \times 0.5)^{0.5} = 0.5$ and $\sigma_o^2 = 0.25$. σ_∞ is taken as 0. Thus $M = \sigma_m^2/\sigma_o^2 = 4\sigma_m^2$. This gives the following values for M:

θ_m(min):	0	1	2	3	4	5	6
M :	1	0.367	0.223	0.0912	0.0333	0.0135	0.00672

This is plotted as lnM against θ_m in figure 6.2 and the best straight line plotted for the equation $\ln M = -k\theta_m$ {from $M = \exp(-k\theta_m)$}.

(a) In a normal distribution 95.4% of the samples fall within the range $\pm 2\sigma$ around the mean. Thus, approximating 95.4% to 95%, the time required for the mixing is when $2\sigma_m$ = 4% of 0.5 = 0.02, i.e. when $\sigma_m = 0.01$. When $\sigma_m = 0.01$, $M = 4 \times (0.01)^2 = 4 \times 10^{-4}$ and $\ln M = -7.824$. From figure 6.2 this gives a mixing time of 9.3 minutes.

(b) From figure 6.2, k = -gradient = 0.844. Since $k \propto N$, an increase in N of 50% increases k to 1.5k, i.e. to 1.266, and hence for lnM = -7.824, θ_m = 6.2 minutes.

(c) If the liquid height in the vessel (H) is proportional to the batch size (further information on the geometry of the mixer would be required for exact evaluation), a reduction of 20% in the

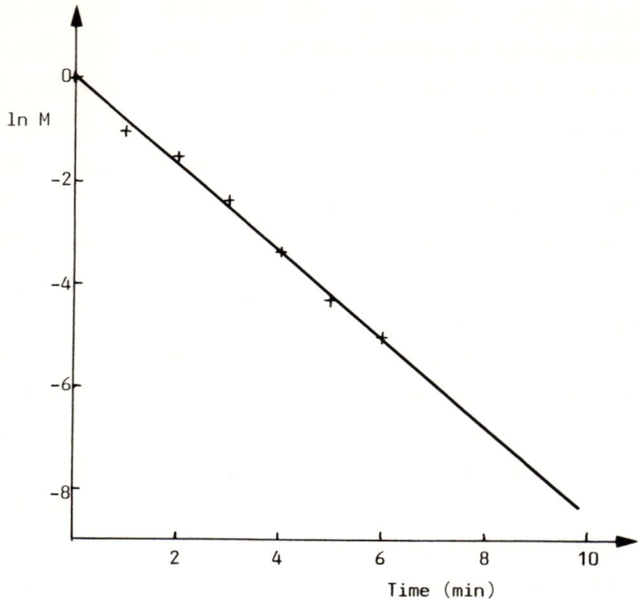

Figure 6.2 - Mixing Index against time for example 6.2

batch size reduces the liquid height to 0.8 of its original value. Since $k \propto 1/H$, k increases to 0.844/0.8 = 1.055. Hence, for lnM = -7.824, Θ_m = 7.4 minutes.

Example 6.4

If the power consumption in a mixer is to be kept constant, what change in the mixing rate constant will be caused by a doubling of agitator diameter? Assume flow conditions in the mixer are in the laminar region (i.e. Re<20) and the liquids to be mixed are Newtonian.

In the laminar region $Po \propto Re^{-1}$ (see figure 6.1), from which $P \propto d_a^3N^2$. (μ is constant for Newtonian liquids). Therefore for P to be constant $d_a^3N^2$ is constant (or $N \propto d_a^{-1.5}$) and a change in agitator diameter would necessitate a change in agitator speed. Since $k \propto d_a^3N/d_t^2H$, $k \propto d_a^3(d_a)^{-1.5}$ for constant d_t and H. Thus $k \propto d_a^{1.5}$ and doubling the agitator diameter would increase k to 2.83k.

Example 6.5

A laboratory scale mixer is used for the blending of two liquids and it is found that the time to achieve a satisfactory degree of mixing is ten minutes. The laboratory mixer has a bowl of 30cm diameter and the bowl is unbaffled. The agitator is a four-bladed paddle-type of diameter 8cm running at 100 r.p.m. The liquids are (a) a Newtonian fluid of viscosity 0.1kg/ms and density $1000 kg/m^3$ and (b) a pseudoplastic of properties as given in example 6.2. The volume ratio of (a) to (b) in the mix is 1:4.

This mixing operation is to be scaled up to a mixer bowl of diameter 120cm. It is required to keep the mixing conditions as similar as possible to the laboratory mixer. Suggest a suitable agitator system for the process.

For the small mixer, the pseudoplastic apparent viscosity is $R/\dot{\gamma}$ and $\dot{\gamma} = 10N = 10 \times 100/60 = 16.7 s^{-1}$. $R = 4 \times 16.7^{0.5} = 16.33$. Therefore apparent viscosity = 16.33/16.7 = 0.980kg/ms. $\mu_m = \mu_1^{v_1}\mu_2^{v_2} = (0.1)^{0.2}(0.980)^{0.8} = 0.621$kg/ms. $\rho_m = 1000 kg/m^3$ and hence $Re = 0.08^2 \times (100/60) \times 1000/0.621 = 17.2$.

One possible design would be to scale-up the agitator diameter in direct proportion to the increase in bowl size, i.e. keep geometrical similarity. The mixing rate constant $k \propto d_a^3 N/d_t^2 H$; therefore if d_a, d_t and H all increase four fold, to keep the same mixing rate requires the same agitator speed in both the small and large mixers. However the flow conditions at the agitator would be very different since $Re \propto d_a^2$ and therefore Re will increase from 17.2 to 275 in the large mixer.

An alternative design would be to attempt to keep both Re <u>and</u> mixing time the same in both mixers. Say that this would give an agitator diameter and speed in the large mixer of d_{aL} and N_L. Then $\mu_m = \{4 \times (10N_L)^{0.5}/10N_L\}^{0.8}(0.1)^{0.2}$ and $Re = 17.2 = (d_{aL}^2)(N_L) \times 1000/\{4 \times (10N_L)^{0.5}/10N_L\}^{0.8}(0.1)^{0.2}$. This simplifies to $d_{aL}^2 N_L^{1.4} = 0.0131$. For equal mixing rates, $d_a^3 N/d_t^2 H$ has to have a constant value. Therefore for d_t and H both increasing four-fold, $d_a^3 N$ has to increase by 4^3 (=64). For the small mixer $d_a^3 N$ is $0.08^3 \times (100/60) = 8.53 \times 10^{-4}$ and therefore $d_{aL}^3 N_L = 0.0546$. Since for constant Re, $d_{aL}^2 N_L^{1.4} = 0.0131$ the values of d_{aL} and N_L can be determined. These are $d_{aL} = 1.13$m and $N_L = 2.3$ r.p.m. It is clear that this design would not be practical since the geometrical similarity has been greatly altered.

A third possibility is therefore to use a larger $d_a:d_t$ ratio than in the small mixer and to accept a somewhat lower mixing rate. Say the $d_a:d_t$ ratio is reduced to 3 (i.e. d_{aL} = 40cm) and the mixing rate constant decreased by 25%. The latter change leads to an increase in d_a^3N by a factor of 64/1.25 = 51.2. Thus $d_{aL}^3N_L$ = $51.2 \times 8.53 \times 10^{-4} = 0.0437$ and hence N_L = 0.682 r.p.s(41 r.p.m.). Calculating Re as previously from N gives Re = 123.

The proposed design would therefore be to use an agitator of diameter 40cm at a speed of 40 r.p.m.

6.2 BLENDING OF SOLIDS

When particulate materials are to be blended together in a mixer it is useful to predict the time required to achieve an acceptable degree of uniformity in the mix. Some preliminary trials are normally necessary to establish mixing characteristics and this section shows how information on mixer performance can be obtained from such trials.

As with liquid mixing, a mixing index (M) can be used to define the extent of 'mixedness' of the batch of material in the mixer, where $M = \exp(-k\Theta_m)$, k being the mixing rate constant and Θ_m the mixing time. A number of mixing indices have been proposed and the suitability of the index depends on the characteristics of the mixer and the materials being blended. Three mixing indices are considered here: $M_1 = (\sigma_m^2-\sigma_\infty^2)/(\sigma_o^2-\sigma_\infty^2)$, $M_2 = (\sigma_m-\sigma_\infty)/(\sigma_o-\sigma_\infty)$ and $M_3 = (\log\sigma_m-\log\sigma_\infty)/(\log\sigma_o-\log\sigma_\infty)$. σ_o, σ_m and σ_∞ have been defined in section 6.1. In solids mixing it is unlikely that complete uniformity can be achieved in the mix (i.e. $\sigma_\infty \neq 0$) and it is therefore necessary to establish a suitable level of best possible mixing. This should be set at a level which gives a value of M significantly different from zero for an acceptably mixed batch.

The use of these indices is shown in the following examples.

Example 6.6

A mixer is used for blending 100kg batches of flour with 600g of a powdered seasoning, i.e. to give a flour containing 0.6% seasoning. The mix is filled into 500g capacity cartons and it is required that there will be no more than a 5% probability that any carton will contain seasoning outside the range 0.57% - 0.63%. Ten 500g samples are taken from the mixer after various mixing times and analysed for % seasoning with the results tabulated below.

Time: 75s		100s		125s		250s		500s	
0.95	0.25	0.46	0.78	0.29	0.70	0.63	0.60	0.62	0.61
1.04	0.15	0.50	0.38	0.57	0.66	0.68	0.62	0.59	0.60
0.88	0.92	0.60	0.84	0.72	0.75	0.54	0.59	0.61	0.62
0.18	0.35	0.27	0.81	0.59	0.45	0.62	0.63	0.60	0.60
0.20	0.48	0.68	0.58	0.67	0.50	0.61	0.58	0.58	0.60

(a) Does the sampling show that the mixing requirements will be met after 500s?

(b) Calculate the mixing indices M_1, M_2 and M_3 for each mixing time and draw conclusions from them.

(a) The mean of the samples ($\bar{x}$) after 500s is 0.603%. $\sigma_m^2 = \{1/(n-1)\} \sum_{i=1}^{n} (x_i-\bar{x})^2$, where x_i is the % seasoning in sample i. For the 500s samples this gives $\sigma_m^2 = 1.567 \times 10^{-4}$ and $\sigma_m = 0.0125\%$. The probability of samples taken from a normal distribution lying within the range mean $\pm$ 2σ is 95.43%. Therefore 95.43% of cartons taken from the mixer will have seasoning concentrations in the range $0.6 \pm 0.025\%$ i.e. between 0.575% and 0.625%. Therefore the mixing requirement is met.

(b) Take as 'perfect mixing' a probability of 99.73% that cartons will fall within the desired range of 0.57% to 0.63%. 99.73% of samples from a normal distribution fall within $\pm$ 3σ of the mean. Thus this 'perfect mixing' standard gives $3\sigma_\infty = 0.03\%$ and thus $\sigma_\infty = 0.01\%$.

$\sigma_o = \{x_1(1-x_1)\}^{0.5}$, where x_1 is the average fraction of one component in the mix. In this case $x_1 = 600/100 \times 10^3 = 6 \times 10^{-3}$ and $\sigma_o = (6 \times 10^{-3} \times 0.994)^{0.5} = 0.07723 = 7.723\%$.

The table below shows the values of M_1, M_2 and M_3 as calculated from $M_1 = (\sigma_m{}^2 - \sigma_\infty{}^2)/(\sigma_o{}^2 - \sigma_\infty{}^2)$, $M_2 = (\sigma_m - \sigma_\infty)/(\sigma_o - \sigma_\infty)$ and $M_3 = (\log\sigma_m - \log\sigma_\infty)/(\log\sigma_o - \log\sigma_\infty)$.

	Time(s)				
	75	100	125	250	500
$\bar{x}$	0.540	0.590	0.590	0.610	0.603
σ_m	0.365	0.191	0.143	0.0368	0.0125
$\sigma_m{}^2$	0.133	3.63×10^{-2}	2.04×10^{-2}	1.36×10^{-3}	1.567×10^{-4}
M_1	2.23×10^{-3}	6.07×10^{-4}	3.41×10^{-4}	2.11×10^{-5}	9.51×10^{-7}
M_2	4.60×10^{-2}	2.34×10^{-2}	1.72×10^{-2}	3.48×10^{-3}	3.24×10^{-4}
M_3	0.541	0.443	0.400	0.196	0.0336

It can be seen that the M_3 index is more appropriate for describing this mixing operation since M_1 and M_2 show a rapid approach to $M = 0$. This is because of the high σ_o value caused by both the low % seasoning in the mix and the rapid mixing achieved - in other circumstances M_1 or M_2 would be a better index. Figure 6.3 shows mixing index against time plots. It can be seen that only M_3 gives a

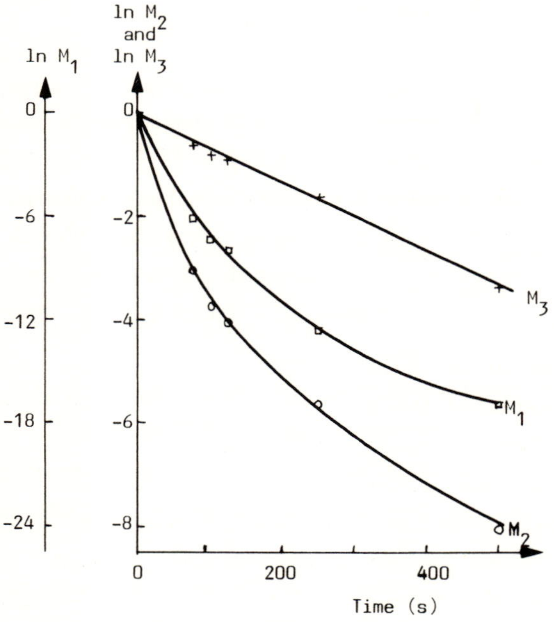

Figure 6.3 - Mixing index against time graphs for example 6.6

relationship of the type $M = \exp(-k\theta_m)$, or $\ln M = -k\theta_m$, which suggests that it gives a better description of the mixing process and also aids interpolation and extrapolation of the data to other mixing times. Notice that there are no significant irregularities in the graphs, which indicates that the mixing system does not suffer from anomalies in the process of mixing (see example 6.7).

Example 6.7

The data shown in figure 6.4 (line 1) were typically obtained in a particular powder mixing operation. The data represented by lines 2 and 3 were obtained for the same process on two further occasions. Comment on the significance of these atypical findings.

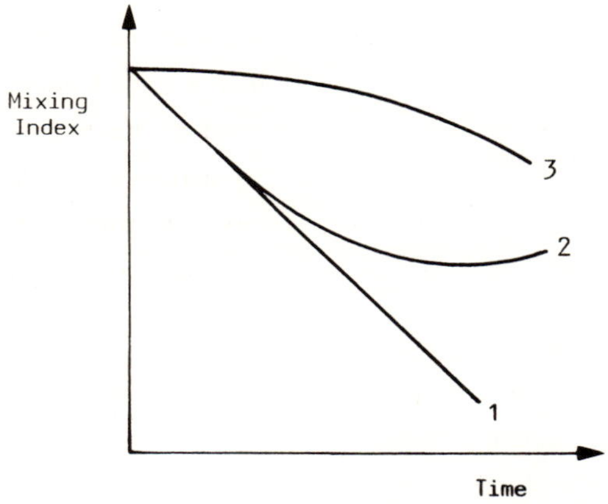

Figure 6.4 - Mixing data for example 6.7

Line 2. There is apparent de-mixing taking place, causing a rise in mixing index at longer mixing times. This is a not uncommon phenomenon in particle mixing systems and arises from one component separating out of the bulk of the mix - it is most likely to be due to the components of the mix having very different sizes, shapes or densities. Possible causes of this mixing behaviour would therefore be that one or more of the ingredients of the mix was different in these physical properties from normal (perhaps an excessively large quantity of 'fines' in one of the powders), or that there was an unusual degree of agglomeration of particles in the mix. This latter effect could be caused by the presence of excessive moisture or fat.

Line 3. An initial lag before the mixing index falls suggests that the initial conditions in the mixer were abnormal, e.g. the components were added in an unusual order, perhaps causing one ingredient to take a longer time than usual to reach the main mixing zone. A useful indication of this is to check the mean % of one ingredient in the set of samples taken at any one time. This should be reasonably close to the actual % of the ingredient in the overall mix. For example the mean% for the samples of example 6.6 at each time interval were 0.540, 0.590, 0.590, 0.610 and 0.603 which are all satisfactorily close to the actual mean of 0.6%.

REFERENCES

1. S.A. Miller and C.A. Mann, 'Agitation of two-phase systems of immiscible liquids', Trans.Am.Inst.chem.Engrs, 40 (1944) 709
2. D.S. Laity and R.E. Treybal, 'Dynamics of liquid agitation in the absence of an air-liquid interface', A.I.Ch.E.Jl, 3 (1957) 176
3. T. Vermeulen *et al*, 'Interfacial area as a measure of agitation', Chem.Engng Prog., 51 (1955) 85F
4. G.G. Brown, *Unit Operations*, (Wiley,New York,1950)
5. V.W. Uhl and J.B. Gray, *Mixing, Theory and Practice*, Vol.1, (Academic Press, London, 1966)
6. J.T. Davies, *Turbulence Phenomena*, (Academic Press, London, 1972)
7. H.A. Leniger and W.A. Beverloo, *Food Process Engineering*, (D. Riedel, Dordrecht, 1975)

7 GAS-SOLID SYSTEMS

7.1 FLUIDISATION

When a fluid passes upwards at low velocity through a bed of particles supported on a perforated screen, the particles remain stationary and the bed is referred to as 'fixed' or 'packed'. As the velocity increases a point is reached where the weight of the bed is just balanced by the upward force exerted by the fluid stream. The bed is now 'fluidised' and as the fluid velocity is further increased the bed becomes more open (a greater voidage) until eventually the particles are conveyed out of the bed in the fluid stream (see Chapter 7.2). Fluidised beds have a number of useful characteristics - heat and mass transfer rates to and within the bed are high and there is good uniformity of treatment of the particles constituting the bed. Kunii and Levenspiel [1] give a comprehensive account of the theory and practice of such systems. Although particles can be fluidised in liquids, only gaseous fluidisation is considered here.

Requirements for fluidisation

Since fluidisation occurs when the weight of particles is balanced by the upward force it follows that the pressure drop from bottom to top of the bed (Δp_b) = bed weight/bed cross-sectional area = $L(1-\varepsilon)(\rho_p-\rho_g)g$, where L is the bed height, ε is the fractional gas volume of the bed (voidage) and ρ_p and ρ_g are the particle and gas densities respectively. Note that the pressure drop remains constant as the gas velocity increases, as does $L(1-\varepsilon)$. Kunii and Levenspiel [1] give the following equation for calculating the gas velocity at the onset of fluidisation (the minimum fluidising velocity u_{mf}):

$$1.75\,(d_p u_{mf}\rho_g/\mu_g)^2/\psi\varepsilon_{mf}^3 + 150(1-\varepsilon_{mf})(d_p u_{mf}\rho_g/\mu_g)/\psi^2\varepsilon_{mf}^3 = d_p^3\rho_g(\rho_p-\rho_g)g/\mu_g^2$$

Where ψ is a particle shape factor for non-spherical particles (= surface area of sphere of same volume as particle/surface area of non-spherical particle), μ_g is the gas viscosity, ε_{mf} is the minimum fluidising voidage and d_p is the particle diameter. u_{mf} is the superficial gas velocity (volumetric flow rate/bed cross-sectional area). Some typical values of ψ are given by Carman [2] and of ε_{mf} by Leva [3]. If these properties are not known, Wen and Yu [4] proposed $1/\psi\varepsilon_{mf}^3 \doteqdot 14$ and $1 - \varepsilon_{mf}/\psi^2\varepsilon_{mf}^3 \doteqdot 11$. For $Re_m = d_p u \rho_g/\mu_g < 20$ the first term of the fluidisation velocity equation can be ignored, and for $Re_m > 1000$ the second term can be ignored.

Example 7.1

Calculate (a) the air velocity required to fluidise 100kg of spherical particles of 5mm diameter, density 1000kg/m^3 in an air stream of density 1kg/m^3, viscosity 1.6×10^{-5}kg/ms, (b) the pressure drop through the bed if the cross-sectional area is 0.3m^2 and (c) the bed height at a voidage 20% above the minimum.

(a) Since no value for ε_{mf} is given, use the Wen and Yu formulae. ψ is 1 for a spherical particle. $d_p u_{mf}\rho_g/\mu_g = Re_m = 312.5u_{mf}$. Thus, $1.75 \times 9.766 \times 10^4 \times u_{mf}^2 \times 14 + 150 \times 11 \times 312.5u_{mf} = 0.005^3 \times 1 \times 999 \times 9.81/(1.6 \times 10^{-5})^2 = 4.785 \times 10^6$. This gives $u_{mf} = 1.3$ m/s.

(b) Pressure drop = $100 \times 9.81/0.3 = 3270$ Pa

(c) $\varepsilon_{mf} = (1/14)^{0.333} = 0.415$. Thus voidage at 20% above minimum is 0.498 and bed height = bed volume/area $=\{100/(1 - 0.498)(1000)\}/0.3 = 0.66$m.

It is important that there is an even distribution of the fluidising gas across the bed and this is aided by causing a sufficient resistance to flow through the distributor plate on which the bed rests. Kunii and Levenspiel [1] recommend that the minimum pressure drop through the plate should be either $0.1\Delta p_b$ or 100 x (pressure drop in the expansion duct into the bed) or 3500Pa, whichever is the largest.

Heat transfer in fluidised beds

Bed wall to bed. Wen and Leva [5] correlated data from a number of investigators fluidising particles up to 0.85mm in diameter, to give the following equation for the heat transfer coefficient between the retaining wall and the bed (h_w):

$$h_w d_p/k_g = 0.16(c_p \rho_p d_p^{1.5} g^{0.5}/k_g)^{0.4}(d_p u_o \rho_g L_{mf} E/\mu_g L)^{0.36}$$

where k_g is the thermal conductivity of the gas, c_p is the particle specific heat, L_{mf} is the bed height at u_{mf} and E is a fluidisation efficiency factor, values for which are shown in figure 7.1.

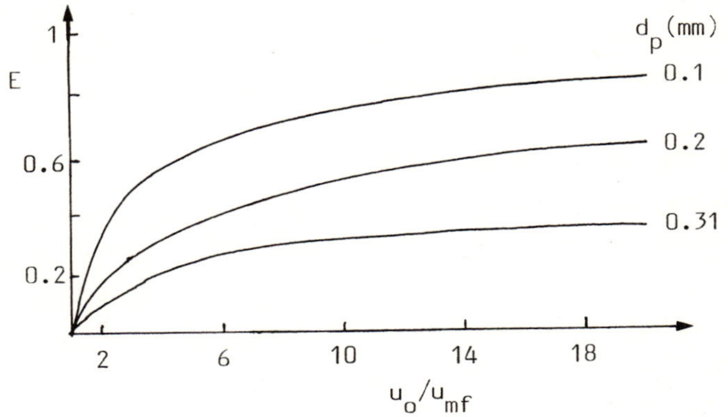

Figure 7.1 - Fluidisation efficiency factor (from Wen and Leva [5])

Levenspiel and Walton [6] working with larger particles (up to 4.3mm) found the correlation $h_w d_p/k_g = 0.6\ (c_g \mu_g/k_g)(d_p \rho_g u_o/\mu_g)^{0.3}$, where c_g is the gas specific heat (at constant pressure).

Since heat transfer between the fluid and particles is normally very rapid (see later) it may be assumed that the particles approach the gas temperature throughout most of the bed in a batch operation. The gas and bed temperature then depends on a heat balance between the heat flowing into the bed from the walls and the heat taken up by the gas and particles. If the gas enters at temperature t_o and exits at t_e after a time θ, then for a wall heating medium temperature t_w and particles initially at t_o, a heat balance gives $dt_e/d\theta = 2(t_e - t_o)\{[UA/\ln(t_w - t_o)/(t_w - t_e)] - m_g c_g\}/m_p c_p$.
Integration of this expression gives t_e at time θ, the maximum temperature achievable being when $dt_e/d\theta = 0$, i.e. when $(t_w - t_o)/(t_w - t_e) = \exp(UA/m_g c_g)$. (U is the overall heat transfer coefficient from heating medium to bed, A is the surface area for heat transfer, m_p and c_p are bed mass and specific heat, m_g and c_g are gas mass flow rate and specific heat).

Example 7.2

The bed of particles of example 7.1 is heated through the bed wall with a heating medium at 150°C and a heat transfer coefficient from medium to wall of $500 W/m^2K$. The air velocity is $3u_{mf}$. The bed wall is 2mm thick and has a thermal conductivity of 40W/mK. If the fluidising air enters the bed at 20°C calculate bed temperature achievable. For the air, c_g = 1050J/kgK, k_g = 0.025W/mK. The particles have a specific heat of 3500J/kgK.

Since the particles are large we will use Levenspiel and Walton's correlation [6]: thus $h_w \times 0.005/0.025 = 0.6(1050 \times 1.6 \times 10^{-5}/0.025)[(0.005 \times 1 \times 3 \times 1.3)/(1.6 \times 10^{-5})]^{0.3}$. From this h_w = $17.0 W/m^2K$. The bed height change with velocity is difficult to predict, but a reasonable figure for a velocity of $3u_{mf}$ would be $1.2L_{mf}$[1, 7]. Thus $A = 1.2 \times L_{mf} \times$ bed diameter $\times \pi = 1.2 \times \{100/(1 - 0.415)(1000)\}/0.3 \times (4 \times 0.3/\pi)^{0.5} \times \pi = 1.328 m^2$. $1/U = (1/500) + (0.002/40) + (1/17)$ and $U = 16.4 W/m^2K$. Hence $(150 - 20)/(150 - t_e) = \exp(16.4 \times 1.328)/(3 \times 1.31 \times 0.3 \times 1)1050 = 1.0177$. Therefore maximum temperature achievable = 22.3°C. It will be seen that the high air flows and comparatively low wall heat transfer coefficients associated with fluidising large particles make heating via the wall impractical.

Example 7.3

Calculate (a) the maximum temperature achievable and (b) the time for the bed to reach a temperature of 80°C, for a bed of particles of 0.5 mm diameter, all other properties being as in example 7.2.

(a) Calculating the minimum fluidising velocity as in example 7.1 gives u_{mf} = 0.0891m/s. Thus air velocity in bed = $3u_{mf}$ = 0.267m/s. Applying the Wen and Leva equation for h_w and taking E = 0.1 from figure 7.1 gives $h_w = 210 W/m^2K$ and thus $U = 147 W/m^2K$. Therefore $130/(150-t_e) = \exp(147 \times 1.328)/(0.267 \times 0.3 \times 1)1050$ for maximum t_e, which gives a maximum bed temperature of 137°C.

(b) Substituting all known properties into the equation for $dt_e/d\theta$ gives the $dt_e/d\theta = -4.806 \times 10^{-4}(t_e-20) + 11.155 \times 10^{-4}(t_e-20) \ln(130/150-t_e)$. Writing the right hand side of this equation as

$f(t_e)$, calculating $1/f(t_e)$ for various values of t_e gives the data tabulated below:

t_e	$1/f(t_e)$	t_e	$1/f(t_e)$
20*	6.90	100	18.2
30	7.43	110	23.9
50	8.84	130	79.2
80	12.6		

*it can be shown that when $t_e = t_o$, $dt_e/d\theta[=f(t_e)] = 2UA(t_w - t_o)/m_p c_p$, which is used to calculate this value.

Since $dt_e/d\theta = f(t_e)$, $\int_{t_o}^{t_e} dt_e/f(t_e) = \int_o^\theta d\theta = \theta$ and hence the area under the graph of $1/f(t_e)$ against t_e, from t_o to t_e, gives the time to temperature t_e. This graph is shown in figure 7.2, the area under the curve from $t_e = 20$ to $t_e = 80$ being 550s. Thus the time for the bed to reach 80°C is 550s.

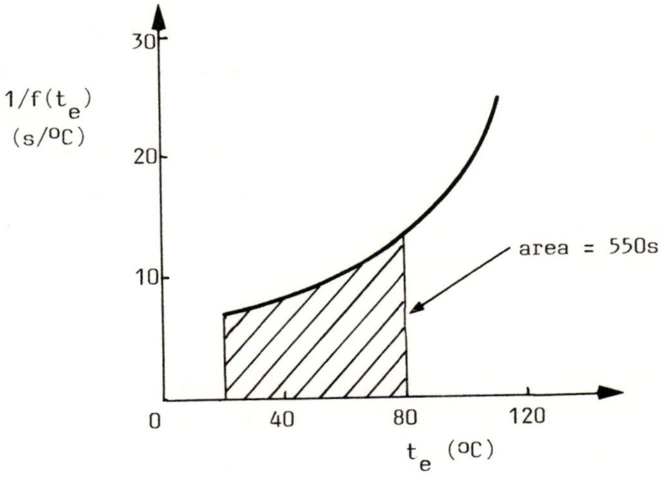

Figure 7.2 - Graph of t_e against $1/f(t_e)$ for example 7.3

Fluidised bed/gas heat transfer. When gas enters a batch fluidised bed at a temperature (t_i) different from that of the particles (initially at t_o), then assuming no heat transfer at the bed wall a heat balance gives [1] : $(t_i - t_e)/(t_i - t_o) = \exp(-m_g c_g\ \theta/m_p c_p)$, where t_e is the temperature of the bed and out-going gas at time θ. For the bed and outgoing gas to be at the same temperature requires that there is a very rapid approach of particle temperature to gas temperature - with the heat transfer coefficients between particle and gas and the small particle mass typical of many fluidisation processes this is the usual situation. Where the enthalpy change in the particles is large e.g. in freezing pieces of foodstuffs, it may be necessary to calculate the heat transfer coefficient between particle and gas (h_p). Kunii and Levenspiel [1] propose the correlation $h_p d_p/k_g = 0.03(d_p u_o \rho_f/\mu_f)^{1.3}$ for calculating h_p for Re_m<100. There is little reported work on heat transfer to larger particles (with higher Re_m): Balakrishnan and Pei [8] proposed $h_p/c_g u_o \rho_g = 0.043[(u_o d_p^{-0.5} g^{-0.5})(\rho_g/\rho_p-\rho_g)^{0.5}(1-\varepsilon)^{-0.5}]^{-0.5}[c_g \mu_g/k_g]^{-0.67}$.

Example 7.4

Peas of density 950kg/m^3 and diameter 8 mm are to be frozen in a fluidised bed freezer, with refrigerated air used as the fluidising medium at a velocity 10% above minimum fluidising velocity. If the bed has a voidage of 0.45, air density = 1.2kg/m^3, air viscosity = 1.8×10^{-5}kg/ms, air thermal conductivity = 0.022W/mK, air specific heat = 1000J/kgK, calculate the heat transfer coefficient between the peas and the air.

Calculating u_{mf} as previously gives u_{mf} = 1.75m/s. Therefore required velocity = 1.75 + 10% = 1.93m/s. $Re_m = d_p u_o \rho_f/\mu_f$ = 1029. This is high for the Kunii and Levenspiel correlation $h_p d_p/k_g = 0.03\,Re_m^{1.3}$, but applying this gives $h_p = 0.03 \times 8244 \times 0.022/0.008$ = 680W/m^2K. The Balakrishnan and Pei correlation gives $h_p/1000 \times 1.93 \times 1.2 = 0.043(6.889 \times 0.0356 \times 1.348)^{-0.5}(0.818)^{-0.67}$, giving h_p = 198W/m^2K. For these fluidising conditions the latter figure would be the more appropriate value to use for the heat transfer coefficient. Chapter 2.4 shows how h_p is used in calculating the freezing time of the product.

Example 7.5

The bed of particles of example 7.1 is heated directly by the fluidising air (cf example 7.2). The air enters at 150°C and its velocity is $3u_{mf}$. Calculate the time to raise the particles temperature from 20°C to 80°C if the air specific heat is 1050J/kgK and the particle specific heat is 3500J/kgK.

$(t_i-t_e)/(t_i-t_o) = \exp(-m_g c_g \theta/m_p c_p)$. Hence $(150-80)/130 = \exp(-1.31 \times 3 \times 1 \times 0.3 \times 1050\theta/100 \times 3500)$, which gives the heating time $\theta = 175$s.

Drying in fluidised beds.

The process of drying has been analysed in Chapter 4.1. When low humidity air is used to fluidise moist particles there will be an initial constant rate period of drying if the particles are above their critical moisture content. Equating heat required for evaporation of water from the bed of particles to heat lost by the air passing through the bed gives moisture loss per unit time per unit particles surface area = $F_1 = A_x \rho_g u_o c_g (t_i-t_e) d_p \rho_p / 6m_p \lambda$, where A_x is the bed cross sectional area, t_i is the inlet air temperature, t_e is the outlet air temperature (which will be the wet-bulb temperature) and λ is the latent heat of evaporation of water. Assuming no particle shrinkage on drying m_p/ρ_p remains constant and therefore in the above equation the mass of the bed at any known particle densitycan be used. This calculation of F_1 assumes no significant change in air humidity as the air passes through the bed. Equating moisture uptake by the air to moisture loss from the bed shows this assumption will hold if $c_g(t_i-t_e)/\lambda \ll H_i$ is the absolute humidity of the inlet air. If this is not the case, the outlet air humidity is calculated as $H_i + c_g(t_i-t_e)/\lambda$ and an average t_e based on the inlet and outlet conditions is determined.

Having obtained F_1 the analysis of the drying time given in Chapter 4.1 can be used, except that the slab thickness L in that analysis is replaced by $d_p/6$, since this is the volume/unit surface area of a sphere.

Example 7.6

The particles in example 7.1 are hygroscopic and non-porous and have a moisture content initially of 200% (dry weight basis) and a critical moisture content of 150%. The bed of these particles is dried in fluidising air entering the bed at 70°C and 25% R.H. at a velocity $3u_{mf}$. At these air conditions the equilibrium moisture content of the particles is 3%. Calculate the time to dry the particles to 8% moisture content. Take c_g = 1050J/kgK, λ = 2.3 x 10^6J/kg and particle diffusion coefficient = 5 x $10^{-8}m^2/s$.

From the psychrometric chart (figure 4.2) the wet-bulb temperature t_e = 47.5°C for air at 70°C, 25% R.H. Therefore F_1 = 0.3 x 1 x 3.93 x 1050 x 22.5 x 0.005 x 1000/6 x 100 x 2.3 x 10^6 = 1.01 x $10^{-4}kg/m^2s$. Checking the rise in humidity of the outlet air, $c_g(t_i-t_e)/\lambda$ = 1050 x 22.5/2.3 x 10^6 = 0.010. H_i = 0.065 from figure 4.2. Thus outlet air humidity = 0.075 and at 70°C this is equivalent to a wet bulb temperature of 49°C. The difference from t_e is small and will be ignored. Volume of particles = mass/density = 100/1000 = $0.1m^3$. Surface area: volume ratio of a sphere = 6/dp. Therefore surface area of all particles in bed = (6/0.005) x 0.1 = $120m^2$. Thus rate of moisture loss from bed in constant rate period = $120F_1$ = 0.012kg/s Initial m.c. of bed = 100 x (2/3) = 66.67kg. Critical moisture content = 33.33 x 1.5 = 50kg. Thus moisture to be removed = 16.67kg. Therefore constant rate drying time = 16.67/0.012 = 1389s = 23.1 min.

The equation for calculating the rate of drying of a non-porous hygroscopic sphere in the second drying period is given in Chapter 4.1 and leads to a drying time in this period of 9.7 min. Therefore total drying time is 33 min.

When fresh material is continuously fed into a fluidised bed, with continuous removal of processed material, it is necessary to know the residence time of individual particles in order to calculate rates of heat or mass transfer. Kunii and Levenspiel [1] describe the analysis of such a system.

7.2 PNEUMATIC CONVEYING

For particles to be conveyed in an air stream the velocity of the air must be sufficiently high to stop particles settling out. In flow through a horizontal pipe the minimum air velocity to stop particles settling to the bottom of the pipe is called the saltation velocity. The equivalent velocity for flow through a vertical pipe is the choking velocity.

Saltation velocity.

The procedure proposed by Zenz [9] is used here.

(a) Calculate $d_p^* = d_p/[3\mu_g{}^2/4g\rho_g(\rho_p-\rho_g)]^{1/3}$. If there are a mixture of particle sizes calculate d_p^* for the largest and smallest particles.

(b) From figure 7.3 calculate u_{ss}^* from d_p^*, and hence $u_{ss} = 0.19\ u$ $[4g\mu_g(\rho_p-\rho_g)/3\rho_g{}^2]^{1/3}(d_t)^{0.4}$, where u_{ss} is the minimum conveying velocity for a single particle and d_t is the pipe diameter in mm.

(c) Calculate n, the gradient of the curve in figure 7.3 at d_p^*. In the case of a size mixture n is the slope of the line joining the values of u_{ss}^* for the smallest and largest particles.

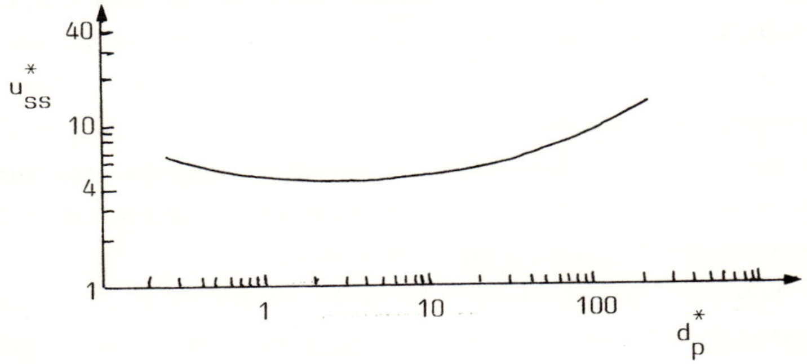

Figure 7.3 - Minimum conveying velocity for a single particle for d_t = 63.5mm (from [9])

(d) Calculate u_s, the saltation velocity, from $G_p/\rho_p = 0.214n^{1.5}$ $(u_s-u_{ss})/u_{ss}$, for $n > 0.068$ or $G_p/\rho_p = 3.2 \times 10^{-3}(u_s-u_{ss})/u_{ss}$ for $-0.11<n<0.068$.

G_p is the solids mass velocity (kg/m^{2}s if u is in m/s). For a size mixture u_{ss} for the largest particle is used in these equations.

Choking velocity.

Kunii and Levenspiel [1] give data on pneumatic conveying that sugges for single size particles that choking velocity (u_c) is approximately equal to saltation velocity and for a size mixture approximately equal to a quarter of the saltation velocity.

Example 7.7

Grain of particle size 2mm and density $1300kg/m^3$ is to be pneumatically conveyed through 7.5cm diameter pipe at a rate of 1.5t/hr. If the air density is $1kg/m^3$ and air viscosity is $2 \times 10^{-5}kg/ms$ calculate the saltation velocity.

(a) $d_p^* = 0.002/[3 \times (2 \times 10^{-5})^2/4 \times 9.81 \times 1 \times 1299]^{1/3} = 69.8$

(b) From figure 7.3, $u_{ss}^* = 8.0$ and $u_{ss} = 0.19 \times 8.0(4 \times 9.81 \times 2 \times 10^{-5} \times 1299/3 \times 1^2)^{1/3}(75)^{0.4} = 5.97m/s$.

(c) From figure 7.3, n(at $d_p^* = 69.8$) = 0.37

(d) G_p = mass flow rate/pipe cross-sectional area = $(1500/3600)/(\pi \times 0.075^2/4) = 94.4kg/m^2s$. Therefore $94.4/1300 = 0.214 \times 0.37^{1.5}(u_s - 5.97)/5.97$, giving $u_s = 15.0m/s$.

Pressure drop in conveying

This is determined by calculating potential energy, kinetic energy and frictional energy changes in horizontal and vertical sections and bends in the conveying line separately.

Horizontal sections. The potential energy loss is zero. The kinetic energy gain is $u_p G_p$ (in absolute units). u_p is the particle velocity and if the particle has accelerated to its maximum value $u_p = u_o - u_t$, where u_o is the air velocity and u_t is the terminal velocity. u_t depends on the flow regime: for $Re_p = d_p \rho_g u_t/\mu_g < 0.4$, $u_t = g(\rho_s-\rho_g)d_p^2/18\mu_g$; for $0.4 < Re_p < 500$, $u_t = \{4(\rho_s-\rho_g)^2g^2/225\rho_g\mu_g\}^{1/3}d_p$; for $500 < Re_p < 2 \times 10^5$, $u_t = \{3.1d_p(\rho_s-\rho_g)g/\rho_g\}^{1/2}$. Since particles are normally fed into a horizontal section it is usual for them to reach a velocity u_p in this section. The frictional energy lost due to the gas flow is $2f_g\rho_g u_o^2 l/d_t$ (in absolute units), where l is the length of the section and f_g is the friction factor. $f_g = 0.0791 Re_g^{-0.25}$ for $3 \times 10^3 < Re_g = \rho_g u_o d_t/\mu_g < 10^5$ or $fg = 0.0008 + 0.0552 Re_g^{-0.237}$ for $10^5 < Re_g < 10^8$. The frictional energy lost due to the particle flow can be estimated [10] as $(\pi/8)(f_p/f_g)(\rho_p/\rho_g)^{1/2}(G_p/G) \times$ frictional energy lost due to the gas flow, where G is the gas mass velocity and f_p is the particle friction factor determined from Re_g as in figure 7.4.

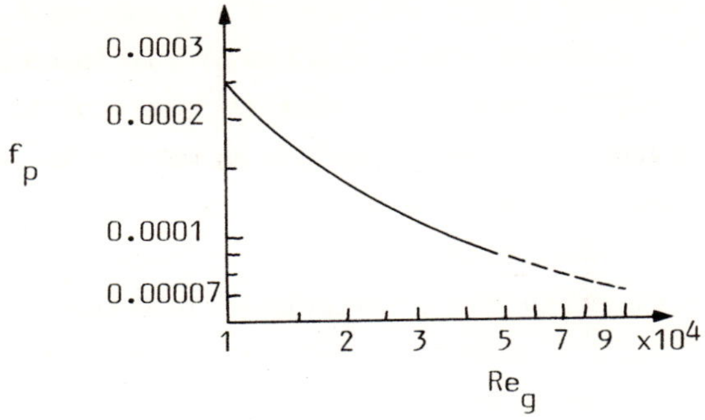

Figure 7.4 - Particle friction factor as function of gas Reynolds Number (from [10])

Vertical sections. The potential energy gain is given by $(G_p/G)(u_o/u_p)h\rho_g g$ (in absolute units), where h is the height of the section. As noted above there will normally be no kinetic energy change in the vertical section. The frictional energy losses are calculated as for the horizontal section.

Bends. The frictional loss in flow round bends can be calculated [11] as $2f_b(G_p/G)(u_o^3/u_p)\rho_g$, where f_b = 0.375, 0.188 or 0.125 for bend radius: pipe diameter ratios of 2, 4 or 6 or more, respectively.

Example 7.8

The pneumatic conveying system of example 7.7 consists of a horizontal section, with 50m straight pipe and two 0.5m radius bends, and a 15m vertical rising section. Calculate the pressure drop through the system if air velocity is 10% above saltation.

Horizontal section: kinetic energy gain = $G_p u_p = 94.4u_p$ N/m^2 (assuming particles reach velocity u_p in this section). u_o = air velocity = 1.1 u_s = 16.5 m/s. To calculate u_t, try the equation for $500 < Re_p < 2 \times 10^5$, giving $u_t = (3.1 \times 0.002 \times 1299 \times 9.81/1)^{1/2}$ = 8.89m/s. Checking Re_p, $Re_p = 0.002 \times 1 \times 8.89/2 \times 10^{-5} = 889$, therefore correct equation used. $u_p = u_o - u_t$ = 7.6m/s. Therefore kinetic energy gain = 94.4 x 7.6 = 717 N/m^2. For frictional losses,

$Re_g = 1 \times 16.5 \times 0.075/2 \times 10^{-5} = 6.19 \times 10^4$, therefore $f_g = 0.0791 (6.19 \times 10^4)^{-0.25} = 0.00502$. Thus, frictional loss due to gas = $2 \times 0.00514 \times 1 \times 16.5^2 \times 50/0.075 = 1821$ N/m^2, $f_p = 0.80 \times 10^{-4}$ from figure 7.4. Thus frictional loss due to particles = $(\pi/8)(0.80 \times 10^{-4}/0.00502)(1300/1)^{1/2}(94.4/16.5 \times 1)(1821) = 2351$ N/m^2.

Vertical section: potential energy gain = $(94.4/16.5 \times 1)(16.5/7.6) \times 15 \times 1 \times 9.81 = 1828$ N/m^2. Frictional loss due to gas = $2 \times 0.00502 \times 1 \times 16.5^2 \times 15/0.075 = 547$ N/m^2. Frictional loss due to particles (calculated as above) = 706 N/m^2.

Bends: radius/pipe diameter = 0.5/0.075 = 6.7. Therefore frictional loss in bends = $2 \times 0.125 \times (94.4/16.5)(16.5^3/7.6) \times 1 = 845$ N/m^2 for each bend. Summing all the above energy losses gives a pressure drop of 8800 N/m^2 in the system.

REFERENCES

1. D. Kunii and O. Levenspiel, Fluidization Engineering, (Wiley, New York, 1969)
2. P.C. Carman, 'Flow through granular beds', Trans.Instn chem. Engrs, 15 (1937) 150
3. M. Leva, Fluidization, (McGraw-Hill, New York, 1959)
4. C.Y. Wen and Y.H. Yu,'A generalized method for predicting the minimum fluidization velocity', A.I.Ch.E.Jl, 12 (1966) 610
5. C.Y. Wen and M. Leva, 'Fluidized-bed heat transfer: a generalized dense -phase correlation', A.I.Ch.E.Jl, 2 (1956) 482
6. O. Levenspiel and J.S. Walton, 'Bed-wall heat transfer in fluidized systems', Chem.Engng Prog.Symp.Ser., 50 (1954) 1
7. F.A. Zenz and D.F. Othmer, Fluidization and Fluid-Particle Systems, (Reinhold, New York, 1960)
8. A.R. Balakrishnan and D.C.T. Pei, 'Fluid-particle heat transfer in gas fluidized beds', Can.J.chem.Engng, 53(2) (1975) 231
9. F.A. Zenz, 'Conveyability of materials of mixed particle size', Ind.Engng Chem. Fundamentals, 3 (1964) 65
10. H.E. Rose and H.E. Barnacle, 'Flow of suspensions of non-cohesive spherical particles in pipes', Engineer,Lond., 203 (1957) 898,939
11. Engineering Equipment Users Association, Pneumatic Handling of Powdered Materials, (Constable, London, 1963)

8 FURTHER EXAMPLES (with answers)

1.1 Sulphuric acid (ρ = 1980 kg/m^3; μ = 26.7 cP) is flowing in a 35 mm bore pipe. If the acid flowrate is 1.0 m^3/minute, what will be the head loss over 30 m of pipe ? (Answer: 285 m of acid)

1.2 Water is pumped to a jacketed vessel situated 15 m above the water storage tank. The cooling water requirement is 4.0 m^3/h, and the diameter of the feed line is 75 mm. The length of the feed line is equivalent to 240 m (including valves and fittings). What will be the head requirement for the circulating pump ? (Answer: 30.7 m)

1.3 A tubular heater consisting of an annulus 50 m long with a 150 mm bore outer tube and 100 mm bore inner tube, is used to keep a salt solution warm by passing the salt solution through the annular space. The rate of flow of the solution is 1500 m^3/day.
(a) What will be the head loss through the heater due to friction ?
(b) If the overall pumping efficiency is 60%, what will be the size of motor required for this duty ? Average density of salt solution = 1150 kg/m^3; average viscosity = 2.3 cP. (Answer: (a) 3.48 m of solution (b) 1140 W)

1.4 A solution of ethanol is pumped to a vessel 25 m above ground level through a 25 mm bore pipe at a rate of 10 m^3/h. The length of pipe is 30 m, and it contains two bends of 20 equivalent diameters each. What is the theoretical head requirement for the pump ? Fluid density = 815 kg/m^3; viscosity = 1.4 x 10^{-3} kg/m s. (Answer: 62.9 m of solution)

1.5 A refrigerated glycol solution is used for cooling a vessel by circulation through half-coils wound round the vessel as a jacket. The coils are of semi-circular cross-section, made from 100 mm diameter rolled steel. The rate of flow of the glycol solution is 30 tonnes/h, and the coils are equivalent to 130 m of straight section. What will be the pressure loss through the coil-jacket ? Density = 1075 kg/m^3; viscosity = 10.3 cP. (Answer: 1.33 bar)

1.6 A power law fluid gave the following results in the laboratory

Shear Stress (N/m^2)	15.5	31.0	67.0	104.0
Shear Rate (s^{-1})	10	30	100	200

It is required to transfer this fluid from a storage vessel at a rate of 15 m^3/h. The length of pipework is 80 m, and the safe working pressure of the pipe is 5.0 bar. What is the smallest size of pipe which can be used ? (Answer: 62 mm bore)

1.7 Laboratory work on a non-Newtonian fluid showed that over the shear rate range 10 to 150 s^{-1}, the value of the fluid consistency (K) was 4.24, and the flow behaviour index (n) was 0.7. The fluid is circulated in a ring main, 150 mm diameter x 250 m total length, and the flowrate is 2.8 m^3/minute. (a) What pressure must the pump develop ? (b) If the flowrate is increased by 10%, what will be the increase in pressure at the pump ? Fluid density = 1580 kg/m^3. (Answer: (a) 9.0 bar (b) 0.62 bar)

1.8 The following results were obtained in experiments using a series of pipes and measuring the pressure drop for various flow-rates

Pipe Size (mm)	Flowrate (l/min)	$\Delta P/L$ (bar/m)
10	4.7	0.3394
10	9.42	0.480
25	29.5	0.0859
25	58.9	0.1214
30	42.0	0.0653
30	84.8	0.0924

The fluid density = 1310 kg/m^3. (a) What are the characteristics of the fluid ? (b) What pump pressure is required to transfer this fluid through (i) a 10 m long x 25 mm bore line, (ii) a 20 m long x 35 mm bore line ? Flowrate = 5m^3/hour. (Answer: (a) pseudoplastic, K = 3.0, n = 0.5 (b) (i)1.44 bar (ii)1.25 bar)

1.9 A condensate receiver on an evaporator plant operates at a vacuum of 25 inches of mercury. A centrifugal pump is used to remove the condensate, and at the working flowrate the minimum NPSH given by the pump manufacturer is 3.0 m. The vapour pressure of the liquid is 6.75×10^3 N/m^2, losses in the suction line and entry port total 1.4 m of liquid. What height above the pump inlet must the working level of condensate be ? Liquid density = 1100 kg/m^3 1 atmosphere = 30 inches of mercury = 1.013×10^5 N/m^2. (Answer: 2.46 m)

1.10 A centrifugal pump over the range of flowrates 5 to 20 m^3/h has a characteristic curve represented by the straight line

$$H = 36 - 1.3Q$$

H = head developed, m; Q = flowrate, m^3/h.

This pump is used to transfer a salt solution to an evaporator, the discharge point being located 15 m above the pump inlet. The pipeline is 35 mm bore, and the equivalent length of pipe and fittings is 48 m. What will be the expected rate of flow and power requirements if the pumping system is 62% efficient ? Fluid density = 1150 kg/m^3; viscosity = 2.3×10^{-3} kg/m s. (Answer: 8.33 m^3/h at 25.2 m head, 1.06 kW)

1.11 A single-acting reciprocating pump has a cylinder 100 mm diameter and a stroke of 230 mm. The suction line is 5 m long x 25 mm diameter, and the level of liquid in the supply tank is 1.0 m above the pump suction. What will be the maximum speed of the pump and rate of discharge to just avoid separation ? Assume that the piston moves with simple harmonic motion. Separation occurs at an absolute head of 10^4 N/m^2; liquid density = 1050 kg/m^3. (Answer: 29.5 rev/min, 3.2 m^3/h)

2.1 The walls of a baking oven are constructed of 7.5 mm thick steel sheets insulated with 50 mm of magnesia and cased in 0.15 mm stainless steel. The total surface area of the oven is 40 m^2. The rate of heat transfer through the surrounding layer of stagnant air is 20 W/m^{2}K. Calculate the rate of heat loss from the oven if the inside temperature is 160°C and the room temperature is 20°C.

Thermal conductivity of steel = 45 W/m K; thermal conductivity of stainless steel = 26 W/m K; thermal conductivity of magnesia = 0.06 W/m K. (Answer: 6340 W)

2.2 1800 l/h of liquid is flowing through a 40 m long pipe, 75 mm bore x 12.5 mm thick wall. The temperature of the liquid entering the pipe is 100°C. The outer surface heat transfer coefficient is 50 W/m^2K and the inner liquid heat transfer coefficient is 450 W/m^2K. If the external temperature is 25°C, calculate the heat loss from the pipe and the outlet temperature of the liquid. Thermal conductivity of the pipe material = 200 W/m K; mean specific heat of the liquid = 3.0 kJ/kg K; mean density of the liquid = 1200 kg/m^3. (Answer: 35.6 kW, 80.2°C)

2.3 In example 2.2 above, if the pipe is insulated with 60 mm thick lagging and the outer surface heat transfer coefficient is 30 W/m^2K, re-calculate the heat loss and exit temperature of the liquid. Thermal conductivity of the insulation = 0.25 W/m K. (Answer: 1.76 kW, 99.0°C)

2.4 5.7 m^3/h of sulphuric acid is heated using a jacketed pipe 30 mm bore from 20°C to 65°C. The jacket is 50 mm diameter and contains hot oil entering at 150°C leaving at 130°C. Calculate (a) the sulphuric acid heat transfer coefficient (b) the oil side heat transfer coefficient (c) the length of jacketed section required. Neglect the resistance of the pipe wall and use the Colburn expression for the calculation of h. For the sulphuric acid - mean thermal conductivity = 0.75 W/m K; mean viscosity = 26.7 cP; mean density = 1980 kg/m^3; mean specific heat = 2.3 kJ/kg K. For the oil - mean thermal conductivity = 0.63 W/m K; mean density 800 kg/m^3; mean specific heat = 3×10^3 J/kg K. (Answer: (a) 2270 W/m^2K (b) 3060 W/m^2K (c) 27.3 m)

2.5 Oil at 10°C is heated in a horizontal pipe 20 m long with a wall surface temperature of 40°C. The flowrate of oil is 450 l/h and the pipe is 50 mm bore. Calculate the temperature of the oil leaving the pipe and the average heat transfer coefficient. Mean density of oil = 800 kg/m^3; mean specific heat of oil = 3.0 kJ/kg K; mean thermal conductivity of oil = 0.13 W/m K. The viscosity of the oil exhibits a linear relationship with temperature; at 10°C = 0.2 P; at 40°C = 0.1 P. Use the Sieder and Tate modified correlation for calculation of the heat transfer coeffiecient. (Answer: 17.6°C, 27.8 W/m^{2}K)

2.6 Cooked mashed potato is cooled on trays in a chilling unit in which refrigerated air at 2°C is blown over the product surface at high velocity. The depth of product in the trays is 30 mm and it has a temperature on entering the unit of 95°C. The product has thermal conductivity 0.37 W/mK, specific heat 3700 J/kgK and density 1000 kg/m^3. Calculate the temperature of the product after 30 minutes cooling (a) at the centre, (b) at a point 6 mm from the surface, assuming negligible surface heat transfer resistance. (Answer: (a) 18.5°C; (b) 11.7°C)

2.7 Calculate the temperatures in example 2.6 if the heat transfer coefficient between air and product is 250 W/m^{2}K. (Answer: (a) 23.5°C; (b) 15.5°C)

2.8 A long bar of square cross-section (5 cm x 5 cm), thermal conductivity 1 W/m K, thermal diffusivity 1 x 10^{-6} m^2/s is heated from 20°C in an oven at 200°C. Calculate the time for the bar to reach a temperature of 180°C (a) at the centre of the bar, (b) at a point 1 cm in from two adjacent major surfaces. Asume a heat transfer coefficient of 80 W/m^{2}K. (Answer: (a) 11 min; (b) 8 min)

2.9 A batch of oil, volume 200 litres, specific heat 2000 J/kg K, density 850 kg/m^3, is heated in a steam-jacketed, agitated, vessel with heating surface area 1.5 m^2. The initial oil temperature is 20°C. (a) What will be the temperature of the oil after 10 minutes if the steam temperature is 130°C, the condensing steam heat transfer coefficient is 10,000 W/m^{2}K and the jacket to oil heat transfer coefficient is 500 W/m^{2}K ? (b) What would this temperature be if the agitator speed were doubled ? (Answer: (a) 99°C; (b) 113°C)

2.10 A power law fluid is flowing in a 35 mm bore pipe at a rate of 750 kg/h. After a sufficient length of pipe to ensure fully-developed flow, the fluid enters a heating section of jacketed pipe. The fluid enters at 25°C and the jacket contains condensing steam at 120°C. Estimate the length of heating section and its configuration to raise the temperature of the fluid to 65°C. Mean specific heat = 2.7 kJ/kg K; mean thermal conductivity = 1.4 W/m K; generalised rheological parameter, n' = 0.5. (Answer: Two 4.0 m long sections in series)

2.11 A Bingham plastic fluid is to be heated in a 30 mm jacketed pipe using 105°C steam. The mass flowrate of the fluid is 500 kg/h. Estimate the length of heating section to raise the temperature of the fluid from 30°C to 50°C. Thermal conductivity = 1.31 W/m K; specific heat = 3.2 kJ/kg K; mean stress at the wall = 16.1 N/m^2.

The yield stress (R_y) as a function of temperature (t°C) is as follows

$$R_y = 9.0 - 0.0375t, \quad R_y \text{ in } N/m^2.$$

(Answer: 3.9 m)

2.12 A power law fluid characterised by the expression

$$K = 0.8(\exp E/RT)^n$$

E/R = 3850 J/mol Kelvin; n = 0.55, is flowing in a 40 mm bore pipe at a rate of 800 kg/h. After a sufficiently long section to attain fully-developed flow, the fluid enters a series of jacketed heating sections containing steam at 115°C. If the fluid enters the heating bank at 30°C, what will be the length and configuration of the heating bank to achieve a bulk outlet temperature of 70°C ? Mean specific heat = 3.0 kJ/kg K; mean thermal conductivity = 1.18 W/m K. (Answer: 2 x 5.65 m sections in series)

2.13 A Bingham plastic fluid flowing at a rate of 500 kg/h is to be heated using a jacketed pipe, inside diameter 40 mm, using hot water at 90°C. The yield stress as a function of temperature is given by

$$R_y = 10.0 - 0.02t \quad \text{where } t = {}^\circ C, \; R_y = N/m^2.$$

The plastic viscosity (μ_p) is also a function of temperature

$\mu_p = 0.2 - 0.001t$ where μ_p = kg/m s.

Estimate the length of heating section required to raise the temperature of the fluid from 25ºC to 55ºC. Thermal conductivity = 1.28 W/m K; mean density = 1200 kg/m^3; specific heat = 3.0×10^3 J/kg K. (Answer: 7.5 m)

2.14 A puree exhibiting power law characteristics is to be heated in a 25 mm bore jacketed pipe 8.0 m long. The puree enters the heating section at 28ºC and a flowrate of 550 kg/h. What will be the bulk outlet temperature for steam temperature of (a) 110ºC (b) 90ºC (c) 150ºC ? The flow characteristics of the puree are

K at 30ºC = 5.13; K at 150ºC = 1.50; n = 0.45.

A plot of ln (K) against tºC is linear over the stated temperature range. Mean thermal conductivity = 0.68 W/m K; mean specific heat = 3.36 kJ/kg K. (Answer: (a) 49.8ºC (b) 44.1ºC (c) 61.9ºC)

2.15 What is the film heat transfer coefficient to a fluid flowing through a plate heat exchanger under the following conditions: fluid density = 1000 kg/m^3; mean viscosity = 1.2×10^{-3} kg/m s; flow rate 2 m^3/hour; thermal conductivity 0.8 W/m K; specific heat 4000J/kg K. The plates are 200 mm wide with gaps of 2 mm; the heat transfer correlation for the plates is $Nu = 0.20\ Re^{0.67} Pr^{0.33} (\mu/\mu_s)^{0.2}$. (Answer: 20,700 W/$m^2$K)

2.16 A plate heat exchanger is required to cool 10 m^3/hour of milk from 80ºC to 10ºC, using chilled water at 4ºC at a flow rate of 50 m^3/hour. The plates have dimensions 1800 mm x 500 mm (effective surface area = 0.7 m^2) with plate gaps of 1.5 mm. How many plates will be required ? (Milk density = 1030 kg/m^3, specific heat 4000 J/kg K; water density = 1000 kg/m^3, specific heat 4200 J/kg K). The two fluids flow countercurrently through the exchanger and the overall heat transfer coefficient is 2000 W/m^2 K. (Answer: 24 plates

2.17 A liquid, flow rate 20 m^3/hour, is to be heated ffom 5º to 70º and then cooled to 10ºC. The operation is performed in a three section plate heat exchanger as shown in figure 8.1.

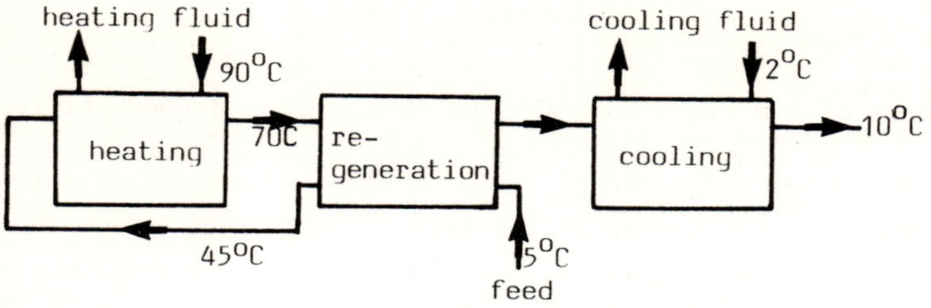

Figure 8.1 - Plate heat exchanger of example 2.17

The heating and cooling liquid flow rates are both 80m^3/hour and they have identical physical properties to the process liquid (specific heat 4000 J/kg K and density 1000 kg/m^3). The overall heat transfer coefficient in the regeneration section is 2000 W/m^{2}K. The plates are 0.8 mm thick, thermal conductivity 16 W/m K, effective surface area 0.7 m^2 and the heat transfer correlation is Nu = 0.20 $Re^{0.65}Pr^{0.35}(\mu/\mu_s)^{0.1}$. Calculate (a) the film heat transfer coefficient for the process fluid, (b) the overall heat transfer coefficient in the heating and cooling section, (c) the outlet temperature of the heating liquid, (d) number of plates in each section. (Answer: (a) 4400 W/m^{2}K, (b) 2700 W/m^{2}K, (c) 83.75°C, (d) heating 10, regeneration 25, cooling 16)

2.18 A meat product packaged in cylindrical casings of diameter 4 cm is frozen in air at -30°C. The latent heat, thermal conductivity (frozen), density (frozen) and specific heat (unfrozen) of the product are 3 x 10^5 J/kg, 1 W/m^{2}K, 800 kg/m^3 and 3000 J/kg K respectively, and the heat transfer coefficient in the freezer is 150 W/m^2 K. Calculate the time to just freeze the product if it enters the freezer (a) at its freezing point of -3°C and (b) at 5°C. (Answer: (a) 24.7 minutes, (b) 28.4 minutes)

2.19 What would (a) air temperature and (b) heat transfer coefficient have to be changed to in example 2.18 to give a freezing time of 20 minutes for the product entering at its freezing point ? (Answer: (a) -36°C, (b) 290 W/m^2 K)

2.20 A food product passes into and through a freezer 10 m in length on a conveyor belt. The product is packaged in cartons 10 cm x 10 cm x 2 cm and these are arranged on the belt so that there are six cartons across the width of the belt and a spacing of 5 cm between cartons along the length of the belt. The product enters at 8°C, leaves at a centre temperature of -15°C, and has a freezing point of -3°C. The thermal conductivity, specific heat and density of the frozen product are 1 W/m K, 1800 J/kg K, 900 kg/m^3, and for the unfrozen product 0.4 W/m K, 3000 J/kg K and 950 kg/m^3 respectively, and the latent heat is 3×10^5 J/kg. The heat transfer coefficient in the freezer is 200 W/m^2K and the freezing air temperature is -30°C. Calculate (a) the conveyor belt speed required and (b) the throughput which can be achieved.
(Answer: (a) 1.1 cm/s (b) 290 kg/hour)

2.21 Use the Neumann equation to calculate the time for a product with the same physical properties as in example 2.20 to be just frozen at a point 3 mm from the surface, using a freezing medium of temperature -30°C. The initial temperature of the product is 8°C.
(Answer: 56s)

2.22 A cold store has dimensions 5 m height, 10 m width and 20 m length and operates at -20°C. Walls and roof are insulated with 100 mm thick foamed polystyrene insulation of thermal conductivity 0.035 W/m K. The floor has a thermal conductivity of 0.08 W/m K and thickness 200 mm. The ground on which the floor is laid is kept at 2°C by heating cables.

The cold store is used to hold 50t of frozen meat; 2t/day goes into the cold store at a temperature of -15°C and an identical quantity of product (at very nearly cold store temperature) is removed per day. Calculate the required thermal refrigeration capacity of the refrigeration equipment for this store given the following data: specific heat of frozen meat = 1700 J/kg K; external ambient temperature = 20°C; lighting has 600 W output for 2 hours per day; two people work in the store, each for 2 hours per day. Assume two air changes per day and that the refrigeration 'on' time is to be 75%. (Answer: 14.7 kW)

2.23 For example 2.22, what extra thermal load would occur if (a) the cold store temperature was reduced to -25°C or (b) the meat entered the store at -10°C, or (c) the insulation on the walls and roof were 150 mm thick ? (Answer: (a) 1.47kW, (b) 200W, (c) -2.33kW)

3.1 The table below shows the centre temperature in a can of a food product sterilised in a steam retort operating at 120°C. Calculate (a) the F_o value of the process and (b) the F_o value if the 'steam - off' time is reduced to 85 minutes. Assume a z value of 10°C.

Time (min)	Temperature (°C)	Time (min)	Temperature (°C)
0(steam on)	92.0	100	116.1
10	93.5	110(steam off, water on)	117.0
20	95.5	115	116.1
30	100.0	120	107.0
40	103.0	122	101.2
50	106.5	125	93.5
60	109.2	130	79.0
70	111.6	135	67.0
80	113.5	140	58.0
90	115.0		

(Answer: (a) 14.4, (b) 4.7)

3.2 Calculate (a) the pseudo initial deficit, (b) the f_H value for the heating section, (c) the j value, (d) the j_c value, for the process data of example 3.1 if the come-up-time is 5 minutes and the cooling water temperature is 20°C. (Answer: (a) 46°C, (b) 90 minutes, (c) 1.64, (d) 1.4)

3.3 Calculate the probability of survival of a micro-organism with D value of 100s at 121.1°C and z value of 10°C at the centre of a can subjected to the original process of example 3.1. (Answer: 1 in 4.4×10^8)

3.4 (a) Calculate the f_H value for the product of example 3.1 in a can of 20% smaller diameter and height; (b) Predict the process time (from steam-on to steam-off) required to obtain the same degree of sterility as in the larger can, assuming the same j-value for both cans. (Answer: (a) 57.6 minutes, (b) 81 minutes)

3.5 The sterilisation cycle for a batch of fermentation medium is as follows

heating from 100°C to 121°C in 40 minutes

holding at 121°C for 30 minutes

cooling from 121°C to 100°C in 26 minutes.

What will be the value of N_0/N for this process based on B.Stearo-thermophilus FS 1518 ? (Answer: 2.94×10^{36})

3.6 In problem 3.5 above, what will be the holding time required at 121°C to give a value for ∇ of 65.5 ? (Answer: 20 minutes)

3.7 Laboratory experiments showed that the following cycle on a pilot plant batch steriliser gives reproducible results

heating from 100°C to 121°C = 6 min

holding at 121°C = 20 min

cooling from 121°C to 100°C = 14 min.

These results are to be scaled-up to a manufacturing unit, but until new steam mains can be installed, the maximum temperature which can be attained is 115°C. Commissioning trials indicate the following

time to heat from 100°C to 115°C = 25 min

time to cool from 115°C to 100°C = 14 min.

(a) what holding time at 115°C will be required to reproduce the laboratory results ? (b) assuming that once the new steam main is installed the rates of heating and cooling will be the same, i.e. heating at 0.6°C/min, cooling at 1.07°C/min, what holding time at 121°C will be required to reproduce the laboratory results ? (Answer: (a) 82 min (b) 12 min)

3.8 A continuous steriliser is designed to operate at a flowrate of 1.0 m^3/h and to raise the temperature of the fluid from a temperature of 25°C to 130°C in 45 s. Cooling of the fluid from 130°C to 100°C will take 15 s. The process requires that the population of B. *Subtilis* FS 5230 in the fluid should be reduced by a factor of 10^{20}. What will be the holding time required at 130°C and the volume of the holding section to achieve this result ? Assume that the sterilising contribution of heating and cooling from 130°C are negligible. (Answer: 8.7 minutes, 0.145 m^3)

3.9 A continuous steriliser operates as follows

time to raise liquid from 100°C to 128°C = 20s

time to cool to 100°C = 9 s.

The holding section capacity is 1.5 m^3, and the process operates satisfactorily with a flowrate of 0.3 m^3/min. It is required to increase the throughput of the plant, and it is suggested that this is achieved by raising the holding temperature to 130°C. What increase in throughput will this achieve ? (Answer: 0.462 m^3/min)

3.10 The liquid throughput in problem 3.9 above can be supplemented by sterilising batches of liquid in a spare vessel 22.5 m^3 in capacity. If heating from 100°C to 121°C takes 20 minutes, cooling from 121°C to 100°C 15 minutes, (a) what will be the holding time required at 121°C to give the same result as the continuous steriliser ? (b) what average increase in throughput will be achieved if emptying and filling of the vessel takes 20 minutes? (Answer: (a) 14.7 minutes, (b) 28%)

4.1 (a) Calculate the initial rate of drying of a slab of non-porous material saturated with water when subjected to an air stream with a velocity of 5 m/s parallel to the material surface. The air has a relative humidity of 10% and a temperature of 80°C. Take air density = 1 kg/m^3 and latent heat of evaporation = 2300 kJ/kg. (Answer: 7.5×10^{-4} kg/m^2 s.)

(b) The material has an initial moisture content of 200% (on a dry weight basis) and an initial thickness and density of 2 mm and 1050 kg/m^3 respectively. Estimate the time required to dry the material to its critical moisture content of 80%, assuming drying takes place from one major surface of the slab only. (Answer: 19 minutes)

4.2 A fibrous material in the form of 5 mm thick sheets is dried in air (from one side only). Calculate the time to dry the slab from 36% to 8% moisture content (dry weight basis) given the following information. Reported information on the product suggests a critical moisture content of 90% and a diffusion resistance coefficient of 4. The dry bulk density of the material is 400kg/m^3. The air has a mass velocity of 6 kg/m^2 s parallel to the product surface and dry and wet bulb temperatures of 80°C and 60°C respectively. Under these conditions the equilibrium moisture content of the material is 4%. Other air properties are: density = 1 kg/m^3, latent heat = 2300 kJ/kg, humid heat = 1300 J/kg K, D_{AB}= 3.7 x 10^{-5} m^2/s. (Answer: 62 minutes)

4.3 Calculate the time for drying a vegetable, of spherical shape and diameter 10 mm, from 130% moisture content to 5% (dry weight basis), in air at 80°C and relative humidity 15%. Assume a constant diffusion coefficient of 5 x 10^{-8} m^2/s, an equilibrium moisture content of 3% and a critical moisture content of 150%. (Answer: 49 minutes)

4.4 (a) What air relative humidity (at 80°C) would be necessary to reduce the drying time in example 4.3 by 10% if the equilibrium moisture content of the vegetable is related to relative humidity in the following manner?

relative humidity (%)	equilibrium moisture content (%)
2.5	1.1
5	1.6
10	2.4
15	3.0
20	3.7

(b) What change in sphere diameter would give a 10% reduction in drying time? (Answer: (a) 8.0%, (b) 5.1% reduction)

4.5 A sheet of fibrous porous material 5 mm thick with a dry bulk density of 400% kg/m^3 is dried on a heated plate kept at 80°C. Air is circulated over the product surface at 0.25 kg/m^2 s, temperature 30°C, relative humidity 30%. The material is known to exhibit a first period of drying ending at 40% moisture content (dry weight basis) and a two part second period with a break in the drying rate curve at 10% moisture content.

Estimate the rate of drying when the moisture content of the material has dropped to 12% if the initial moisture content is 40%. The thermal conductivities of the moist and dry materials are 0.30 and 0.05 W/m K respectively. Use a humid heat of 1300 J/kg K, a latent heat of 2300 kJ/kg and a Biot Number of 20. (Answer: 3.2×10^{-4} kg/m^2 s)

4.6 A food product in slice form is to be freeze dried from an initial moisture content of 150% (dry weight basis) using a vacuum chamber at a pressure of 50 Pa and a temperature of 60°C. The thermal conductivity of the dry porous material is 0.03 W/m K, its permeability is 3×10^{-8} kg/s m Pa and its density is 400kg/m^3. Take the latent heat of sublimation as 3×10^6 J/kg. What is (a) the ice-front temperature in the slice during drying and (b) the maximum thickness of slice which could be dried to 4% moisture content in a time of three hours? (Answer: (a) -22.7^{o}C, (b) 11.1 mm)

4.7 In example 4.6, the product is heated in the freeze drier by placing it on a plate at a temperature of 60°C. The frozen material thermal conductivity is 1 W/m K and the heat transfer coefficient at the product surface is 100 W/m^2 K. Calculate the moisture content achieved in a 4 mm thick slice after 15 minutes drying. (Answer: 32%)

4.8 A salt solution is boiled using a stainless steel surface. The boiling point of the solution is 108°C. Calculate the peak heat flux using Kutateladze's correlation. Latent heat of vaporisation = 2.2×10^6 J/kg; liquid density = 1150 kg/m^3; vapour density = 0.6 kg/m^3; surface tension of the liquid at 108°C = 0.057 N/m. (Answer: 1.4×10^6 W/m^2)

4.9 Re-calculate the peak heat flux in problem 4.8 above using the Rohsenow and Griffith expression. (Answer: 1.5×10^6 W/m^2)

4.10 The salt solution in problem 4.8 above is subjected to a constant heat flux of 10^6 W/m^2. Calculate the temperature of the surface required to sustain this heat flux at atmospheric pressure using (a) McNelly's correlation (b) Kutateladze's correlation. Liquid viscosity = 1.2×10^{-3} kg/m s; liquid thermal conductivity = 0.5 W/m K; liquid specific heat = 3.4×10^3 J/kg K. (Answer: (a) 146°C (b) 152°C)

4.11 5000 kg/h of the salt solution in problems 4.8 to 4.10 is to be evaporated using ten 40 mm bore tubes. The liquid enters the tubes at 100°C with a velocity of 0.15 m/s and the evaporation rate is 25%. Calculate the combined convection/boiling heat transfer coefficient using the Davis and David correlation. (Answer: 9660 W/m^{2}K)

4.12 In problem 4.11 above, the tube length is 2.0 m. If the heat transfer coefficient for convection is 1000 W/m^{2}K, calculate (a) the total area of tubes available for combined convection/boiling (b) the combined heat transfer coefficient using Kutateladze's expression. (Answer: (a) 1.7 m^2 (b) 10 800 W/m^{2}K)

4.13 For the evaporation duty in problem 4.11 above, using Gilmour's correlation, calculate the surface temperature required for evaporation. (Answer: 126 °C)

4.14 A climbing film evaporator of tube length 3 m and internal diameter 25 mm is used for evaporating a 10% feed solution. The feed rate is 150 kg/hr and the solution has a density of 1100 kg/m^3, specific heat 4000J/kg K, viscosity 0.002 kg/m s and thermal conductivity 1W/m^2 K. The feed enters at its boiling point of 50°C and the process steam used for heating is at 125°C. The vapour density and viscosity are 0.05 kg/m^3 and 1×10^{-5} kg/m s respectively. Calculate the solution concentration after one pass through the evaporator, assuming a boiling side heat transfer coefficient given by the equation $h_m = 3.5 h_l (1/X_{tt})^{0.5}$ and a steam-side heat transfer coefficient of 8000 W/m^2 K. (Answer: 17.6%)

4.15 In example 4.14, calculate (a) the concentration achieved if the product from the first pass is recirculated through the evaporator, using the same feed rate of 150 kg/hour, (b) the concentration after one pass if the steam temperature is increased to 140°C. (Answer: (a) 31.1%, (b) 23.1%)

4.16 Calculate the maximum solar heat gain through a wall 50 m long x 5.0 m high containing 4 windows, 2 m x 2 m, facing NW at a latitude of 40°S. The wall is constructed of 105 mm thick solid brick, unplastered. (Answer: 37.4 kW)

4.17 Calculate the maximum solar heat gain to a building with a ground area 30 m x 30 m, walls 5.5 m high. Each wall contains two windows (single glass), 1.5 m x 1.0 m, and the diagonal of the building runs N - S (i.e. one wall faces NW). The walls are constructed of 260 mm thick cavity brick with 16 mm plaster, the roof is flat asphalt on 150 mm concrete with 16 mm plaster. The building is at latitude 48°N. (Answer: 96.8 kW)

4.18 For the maximum solar heat gain in problem 4.17 above, calculate the simple ventilation requirements to maintain a room temperature of 24°C if the outside shade temperature is 19°C. Mean specific heat of air = 1.1 kJ/kg K; mean specific volume of the air = 0.8 m^3/kg. (Answer: 14.1 m^3/s)

4.19 The building in problem 4.17 above is to contain 5 people engaged on light bench work. Electric lighting is installed totalling 1500 W, and there will be 10 x 0.5kW motors running throughout the day. Air changes due to stock movement through the door are equivalent to ¼ change/h. If the building is to be maintained at a controlled temperature of 20°C with a maximum shade temperature of 21°C, calculate (a) the maximum conduction gains to the building (b) the air circulation rate to the air conditioning plant using an 'approach temperature' of 5°. Assume a constant ground temperature of 15°C; the floor is 200 mm thick cast concrete. (Answer: (a) 7.78 kW loss (b) 13.0 m^3/s)

5.1 During the filtration of a suspension at a constant pressure of 2.0 bar, 50 l of filtrate were collected in 10 min and 80 l in 20 min. Calculate the time required to collect 120 l of filtrate from this equipment. (Answer: 38 min)

5.2 If the cake in problem 5.2 above is washed with 120 l of water after collecting 120 l of filtrate, estimate the time for washing if the washing pressure is 5.0 bar. Viscosity of filtrate = 2.0 cP; viscosity of water = 1.0 cP. (Answer: simple washing - 12.4 min, thorough washing - 44 min)

5.3 A rotary vacuum filter, 1.0 m diameter x 2.0 m long, rotates at a speed of 0.2 rev/min and handles 40 m^3 of suspension in 24 h. The solids content of the suspension is 20% w/w and the operating vacuum is 15 inches mercury. The voidage of the cake is found to be 0.5 and the cake moisture content 15% w/w. At any one instant 20% of the drum surface is in contact with the suspension.

If the concentration of the suspension is changed to 35% w/w, what effect on the filter capacity will be observed in terms of suspension handled and solid produced, assuming that all other operating conditions remain constant (drum speed, vacuum, immersion, cake moisture content etc)? Neglect the resistance of the filter cloth. Density of filtrate = 1100 kg/m^3; density of 20% suspension = 1260 kg/m^3; density of 35% suspension = 1400 kg/m^3; density of solid = 3000 kg/m^3; viscosity of filtrate = 1.2 cP. (Answer: suspension - 31.1 m^3/24 h, solid - 15 200 kg/24 h)

5.4 A plate and frame filter press is operated to give a constant rate of filtrate of 0.5 m^3/min for 10 min, by which time the pressure has built up to 4.0 bar. The press is then operated under constant pressure conditions until the chambers are filled. The chambers of the press are 1.5 m square x 100 mm thick and there are 20 chambers. In the laboratory 0.2 ml/s of filtrate flowed through a 10 cm diameter x 1.0 cm thick cake at a pressure of 1.0 bar. (a) How long will it take to fill the press during the constant pressure part of the operation ? (b) What will be the total filtrate collected during the whole cycle ? Density of the filtrate = 1000 kg/m^3;

density of the solid = 2200 kg/m^3; viscosity of filtrate = 1.1 cP. (Answer: (a) 98.3 min (b) 22.7 m^3)

5.5 A 20% w/w solid slurry is filtered using a rotary vacuum filter, 1.0 m diameter x 1.5 m long. The filtrate rate with a speed of 1 rev/3 min and a vacuum of 23 inches mercury is 2.0 m^3/h. The cake is found to have a voidage of 0.4 and a moisture content of 10% w/w. At any one time 25% of the drum surface is immersed in the slurry. Extra capacity of 1.0 m^3/h of filtrate is required, and a plate and frame press is available having chambers of 0.4 m x 0.3 m with a wide range of thicknesses. Assuming a pressure of 5×10^5 N/m^2 to be used, and the time taken to discharge and re-make the press to be 60 s per chamber, calculate the minimum number of plates and frames required. Density of filtrate = 900 kg/m^3; density of the solid = 2100 kg/m^3; viscosity of filtrate = 1.0 cP. (Answer: chambers = 5, frame thickness = 51 mm)

5.6 A simple bowl centrifuge 0.25 m diameter x 0.2 high revolves at 3000 rev/min and is used to clarify dirty oil. The smallest particle contaminating the oil is 12 μm, and the solids are allowed to build up to a thickness of 0.02 m at the wall before discharging. If the bowl is operated half-filled with oil, what will be the maximum possible flowrate to ensure clarification of the oil under all conditions. Density of the oil = 800 kg/m^3; density of particles = 1150 kg/m^3; viscosity of the oil = 0.1 kg/m s. Assume that the particle motion is in the laminar region. (Answer: 3.9 l/min)

5.7 In problem 5.6 above, check that a simple tubular machine is the correct machine for this application. (Answer: a disc machine should be used)

5.8 The operation in problem 5.6 above is to be carried out using a disc machine, outer bowl radius 200 mm, inner liquid radius 50 mm, disc cone half angle 45°. The flowrate will be 7.0 l/min and the centrifuge is to be operated at 1500 rev/min. Determine the number of discs required. (Answer: 12)

5.9 A bowl centrifuge 0.2 m outer diameter revolving at 2000 rev/min is to be used to separate the immiscible phases from a liquid/liquid extraction process. The volume of light phase as a proportion of the total volume is 20%. If the inner liquid radius is fixed at 5 mm, what must be the position of the overflow to obtain complete separation ? Density of light phase = 850 kg/m^3; density of heavy phase = 1050 kg/m^3. (Answer: 20 mm radius)

5.10 The following sieve analysis was obtained after grinding a dried foodstuff

Aperture size range (mm)	% retained
0.95 - 0.42	3.25
0.42 - 0.25	14.40
0.25 - 0.152	49.50
0.152 - 0.125	14.90
0.125 - 0.104	0.70
0.104 - 0.075	9.90
0.075 - 0.053	5.10
0.053 - 0.029	2.25

Calculate (a) the weight mean diameter (b) the surface mean diameter of the powder. (Answer: (a) 0.20 mm (b) 0.15 mm)

5.11 A particle count analysis on a powder is given below

Average diameter (mm)	Count
0.05	300
0.20	750
0.35	280
0.65	62
0.80	5

Calculate the weight and surface mean diameters of the powder. (Answer: weight mean = 0.49 mm, surface mean = 0.40 mm)

5.12 On analysis of a polymer powder, the cumulative weight distribution was found to follow a straight line relationship with coordinates as follows

Cumulative weight (%)	Size (μm)
0.0	1.0
100.0	20.0

Calculate (a) the weight mean diameter (b) the surface mean diameter of the polymer. (Answer: (a) 10.5 μm (b) 6.34 μm)

5.13 A ground oilseed containing 25% w/w of oil is contacted in three stages with a liquid solvent. The plant treats 1000 kg batches of oilseed and 400 kg of solvent is used at each stage. Calculate the oil recovery per batch given that the quantity of solution retained by the separated solids is 0.4 kg/kg solids. (Answer: 91.5 %)

5.14 500 kg of a copper ore containing 12% w/w copper sulphate, 3% water and 85% inerts is extracted using 3000 kg of water. The quantity of copper sulphate solution retained in the inert solids is 0.8 kg/kg solids. Calculate (a) the composition of the overflow and underflow (b) the quantity of overflow and underflow (c) the percentage of copper sulphate extracted. (Answer: (a) overflow = 1.95%, underflow = 0.87% (b) overflow = 2735 kg, underflow = 765 kg (c) 88.8%)

5.15 A sugar beet mash contains 14% w/w sugar, 40% water and 46% fibrous inert materials. Experimental work shows that each kg of inert material retains 2.5 kg of solution over all concentration ranges. Calculate (a) the quantity of water required for extraction (b) the number of stages required to extract 95% of the sugar in the form of a solution, concentration 0.16 kg sugar/kg solution for the continuous counter-current treatment of 20 tonnes/day of beet mash. (Answer: (a) 28.8 tonnes/day (b) 9)

5.16 A continuous counter-current multiple stage system treats 1000 kg/h of oilcake containing 32% w/w of oil using 2000 kg/h of recovered iso-propyl ether (oil content 2% w/w). The final overflow rate is 1800 kg/h with an oil content of 18% w/w, and the final underflow composition is 3% w/w oil, 40.3% ether, the rest being inert solids. Calculate the number of stages required. (Answer: 2)

5.17 A mineral containing 6% w/w water, 12% soluble salt, 82% inerts is continuously extracted at the rate of 500 kg/h using 400 kg/h of water. The solution retained by the solid is 0.4 kg/kg and is constant for all solution concentrations. Analysis of the underflow showed a salt concentration 2% w/w. Calculate the proportion of salt recovered. (Answer: 80.9%)

6.1 Equal volumes of two Newtonian liquids are blended in an unbaffled tank of 1.5 m diameter, the agitator of which has a two-bladed paddle of blade width 20 cm and overall length 1.4 m. The liquids being mixed have viscosities of 0.02 and 0.06 kg/ms and both have a density of 1000 kg/m^3.

Calculate the mixer power consumption at paddle speeds of (a) 20 r.p.m. and (b) 60 r.p.m. (Answer: (a) 40W, (b) 860W)

6.2 What is the maximum speed at which the mixer of example 6.1 should be operated if the agitator motor is restricted to supplying a maximum of 3kW? (Answer: 95 r.p.m.)

6.3 Estimate the power consumed in mixing 0.5 m^3 of a pseudoplastic liquid with 1.5 m^3 of a Newtonian liquid (viscosity 0.1 kg/ms, density 800 kg/m^3) in the vessel of example 6.1 operating at a stirrer speed of 60 r.p.m. The pseudoplastic has shear stress (R): shear rate ($\dot{\gamma}$) characteristics given by $R = 5(\dot{\gamma})^{0.4}$ and density 800 kg/m^3. (Answer: 1300W)

6.4 In the mixing of two liquids in a pilot plant mixer it is found that an acceptable degree of mixing is obtained in a 10 min process. The pilot plant is 50 cm diameter and has an agitator of 30 cm diameter, rotating at 100 r.p.m.; the liquid height in the mixer is 40 cm. (a) Estimate the speed of the agitator to be used in a similar design of mixer of diameter 1.5 m, agitator diameter 90 cm and liquid height 120 cm, if the same mixing rate is to be achieved. (b) If k, the mixing rate constant defined in Chapter 6.1, is 0.5 min^{-1}, estimate the time required to reduce the standard deviation of % of one ingredient in samples taken from the mixer to half its value at 10 minutes. (Answer: (a) 100 r.p.m., (b) 13 min)

6.5 Samples taken from a powder mixer after various mixing times show the following standard deviations in the fraction of one ingredient in the second:

Time (θ)	:	50s	100s	200s	400s	600s
Standard deviation (σ):		0.161	0.108	0.042	0.0084	0.0015

The fraction of the analysed ingredient in the mixture is 0.1 and the mix is considered 'perfectly mixed' when analysed samples show a standard deviation of 1×10^{-4}. Which of the three mixing indices given in Chapter 6.2 gives the best correlation with $M = \exp(-k\theta)$, where k is the mixing rate constant and M is the mixing index? (Answer: M_1 and M_2 are both equally satisfactory)

6.6 Use mixing index $M_2 = (\sigma_m - \sigma_\infty)/(\sigma_o - \sigma_\infty)$ to predict the mixing time required to obtain a standard deviation amongst analysed samples of 5×10^{-4} in the mixer of example 6.5 (Answer: 730s)

6.7 When the mixing of example 6.5 is performed with a fraction of the analysed ingredient in the mix of 0.05, it is found that the mixing characteristics are identical to those with a fraction of 0.1, i.e. the same mixing rate constant is achieved. Predict the standard deviation in the analysed samples which would be found after 300 s with this mix. (Answer: 0.014)

7.1 Calculate (a) the air velocity required to fluidise 50 kg of spherical particles of 3mm diameter, density 700 kg/m^3 in an air stream of density 1kg/m^3, viscosity 1.6×10^{-5} kg/ms; (b) the pressure drop if the particles are contained in a 40 cm diameter vessel and (c) the bed height at a voidage 20% above the minimum fluidising voidage of 0.40. (Answer: (a) 0.74 m/s, (b) 3900 Pa, (c) 1.1 m)

7.2 The bed of particles in example 7.1 is fluidised in air with a velocity of 1.5 m/s. The bed is heated through the bed wall with a heating medium at 150°C and a heat transfer coefficient from medium to wall of 100 W/m^{2}K. The wall is made of 3mm thick steel plate of thermal conductivity 40W/mK. Calculate the overall heat transfer coefficient from heating medium to bed. The air has density 1 kg/m^3, viscosity 1.6×10^{-5} kg/m s, specific heat 1050 J/kgK and thermal conductivity 0.025 W/mK. (Answer: 15 W/m^{2}K)

7.3 Under the conditions of example 7.2 the bed voidage is 0.44. Calculate (a) the maximum temperature achieved by the particles if their initial temperature is 20°C, (b) the time to reach a bed temperature of 28°C if the specific heat of the particles is 3000 J/kgK. (Answer: (a) 32°C, (b) 390 s)

7.4 The bed of particles in example 7.1 is heated by the fluidising air stream. Use (a) the Kunii and Levenspiel correlation and (b) the Balakrishnan and Pei correlation given in Chapter 7.1 to calculate the air to bed heat transfer coefficient at minimum fluidising velocity. The air specific heat in 1050 J/kgK and air thermal conductivity = 0.025 W/mK. (Answer: (a) 150 W/m^2K, (b) 95 W/m^2K)

7.5 The particles in example 7.1 have an initial moisture content of 100% (dry weight basis) and a critical moisture content of 80%. The bed is fluidised in air entering at 80°C with a relative humidity of 20%. The air specific heat is 1050 J/kgK, density 1 kg/m^3 and latent heat of evaporation is 2.3×10^6 J/kg. Calculate (a) the rate of drying in the constant rate period, (b) the time to reduce the moisture content to 80% and (c) the humidity of the exit air. (Answer: (a) 10.6×10^{-6} kg/m^2s, (b) 55 min, (c) 0.119 kg/kg)

7.6 Estimate the saltation velocity of spherical particles, with a size range of 1 mm to 2 mm diameter and density 1200 kg/m^3, conveyed through 10 cm diameter pipe at a rate of 2t/hour. The air density is 1 kg/m^3 and air viscosity is 1.5×10^{-5} kg/m s. (Answer: 13 m/s)

7.7 In example 7.6 the pressure drop on conveying is not to exceed 3000 N/m^2. Calculate the maximum allowable horizontal pipe length, if the air velocity is 16 m/s. (Base calculation on maximum particle size) (Answer: 56 m)

7.8 What pressure drop would result from conveying the particles of example 7.6 in air at velocity 16 m/s through a vertical section of piping of height 10 m ? (Answer: 1370 N/m^2)